WE NEED TO CHANGE TO SOLVE THE WATER CRISIS

Cees Buisman

We Need to Change to Solve the Water Crisis

Published in 2018 as *Humanity is not a Plague: How 10 Billion People can Exist Together*

ESSAY

Orginally published by
© Noordboek, Gorredijk, the Netherlands / Cees Buisman 2018

Co-published by IWA Publishing
Alliance House
12 Caxton Street
London SW1H 0QS, UK
Telephone: +44 (0)20 7654 5500
Fax: +44 (0)20 7654 5555
Email: publications@iwap.co.uk
Web: www.iwapublishing.com

ISBN: 9781789061390 (paperback)
ISBN: 9781789061406 (eBook)
ISBN: 9781789061413 (ePUB)

There is enough for everybody's need,
but not for everybody's greed.

Gandhi

1

*Avarice and usury and precaution
must be our gods for a little longer still.
For only they can lead us out of the tunnel
of economic necessity into daylight.*

Lord Keynes (reaction on the crisis in 1930)

Introduction

Nowadays, in our Western societies the most important purpose in life seems to be to live as long and healthy as possible – or rather, to become immortal. This is naturally supposed to happen in luxury with maximum personal prosperity, all with the hopes of achieving happiness. However, these goals are only accessible if we close our eyes to the consequences of such a lifestyle, namely an ever-growing environmental crisis, an increasing gap between the rich and poor parts of the world and a growing feeling of meaninglessness as experienced by an increasing number of people. Strangely enough, the so-called overpopulation (and particularly the growth in poorer countries) is blamed for these global problems, instead of acknowledging that they are the result of our own goals and behavior.

How can it be that we are so focused on ourselves? Could it be that the 'demystification' of the world through science incited this attitude? Science has burdened society in recent years with a number of visions that, in my opinion, are questionable. Not only are we nowadays apparently just a cosmic accident, but we are also solely our brains. Consciousness, or our own free will, appears to be extinct. Man just follows a kind of computer program that is fixed in their genes. Indeed, human intelligence is not even linked to one's consciousness, according to neuroscientists. The ultimate consequence is that human intelligence could very well be replaced by computer intelligence. In my view, this rational scientific vision could lead to major problems, especially in relation to the challenges our planet and humanity are facing.

For example, there is the threat of a climate crisis, a raw materials deadlock and a water scarcity crisis in a world that will soon have 10 billion people. There is a huge increase in the speed with which animal and plant species become extinct. Can these problems be solved with our current science and technology? In this book I will illustrate that the amount of fresh water available to humanity is already fully used and the effects of economic growth on this water scarcity.

Returning to the demystification or rationalization of the world. The main force behind this demystification is that science has given all kinds of descriptions and predictions for natural phenomena. The Theory of Evolution, in particular, undermined the authority of the Churches, whereas along with the massive increase in prosperity (also largely thanks to science) the authority of science grew steadily.

Religion and mystery often gave way to ideas and apparent scientific absolutes. Some scientists think that ultimately there will be an end to mystery. This view is a misapprehension. We have no idea where our universe came from and it is generally acknowledged that we, as humanity, will never be able to look beyond the Big Bang. The origin of the universe, therefore, remains a mystery. Additionally, the emergence of life on Earth is still a great enigma. In other words, the almighty status of science is open to a good deal of debate. In this respect, I will elaborate on how deeply the ideas and authority of science have penetrated our thinking.

What is required is a new vision: a vision that recognizes the consciousness of people, a vision that values people more than what they can produce. Such a vision is a source of hope for our future. The consciousness of mankind has been evolving for 50,000 years to become less self-centered. It is often thought that the ancient peoples from prehistoric times had a higher level of consciousness. That is not the case: in general, they were violent people, according to our current standards. Back then, man's chances of dying from war-related causes was around 50%. Now that chance is in the single digits.

Consciousness is an invisible phenomenon that is often ignored or forgotten, whereas it is decisive for the behavior of a civilization. Although those with a new economic system or technological innovation hope to solve the world problems, this will be impossible without further growth of our consciousness. Namely, the most important characteristic of a growing consciousness is no longer putting our own interest first. We become less self-centered and, in my opinion, better people. The culture of a society is determined by the average of the individual level of consciousness. This has increased enormously in 50,000 years and we are now at a very high level of consciousness. Which is, of course, good news. At the same time, however, the recognition that even more consciousness is called for is further away than ever. Even when less egocentricity is the solution for virtually all disasters that threaten us. A new vision for our society in response to this, is to embrace our consciousness again.

It is becoming clear that science and technology, without ethical consciousness, are slowly but surely turning from a blessing into a danger to mankind. Technologies are being developed to earn money and no longer to solve social problems. They also have a larger impact and are more dangerous, as nuclear disasters have shown us. Additionally, many of the proposed solutions are, on the whole, only beneficial for the rich part of the world, such as underground CO_2 storage. We should focus on technologies that can be used by everyone and that do not cause any further damage to the environment. In this book I am making a case for this nature-based technology. By nature-based technology I mean an approach that recognizes that we cannot understand everything, and that more caution and care is required to deal with nature and ourselves.

The idea that we will be on Earth with 10 billion people in a few years shocks us. But why should it? Is it because of the people who see humanity as a plague and consider us as the greatest threat to biodiversity? Or is it the fear of shortages and that there will soon be too few resources for everyone? Or perhaps because of the vision that the universe – and therefore also the people – are created randomly by colliding particles, and thus we in fact count for little and do not serve a special purpose, as is implicitly expressed by many scientists? These beliefs are instilled in us. Today, every nature film ends with an accusation to humanity. We hear that the extinction of animals is directly related to a reduced space for nature due to the growth of cities and agriculture, alongside the overfishing of the oceans and pollution of the environment. The lack of raw materials, such as oil and food, is also attributed to overpopulation. A theme in many films and books is that humanity is hit by a major disaster, after which only a select number of people can rebuild a new society. We also see fantasies about space travel to get raw materials from planets or, even more futuristically, people colonizing outer space. Furthermore, replacing human intelligence with computer intelligence could cause billions of jobs to disappear. Billions of people then become 'worthless', which will give a whole new meaning to the term 'overpopulation'.

Slum in Kenya (unknown photographer, source: Aqua for All)

From a philosophical point of view, the popularity of ideas such as genetic manipulation, space travel and artificial intelligence are interesting. Insights that emphasize control over mysteries, such as the scientific explanation for the origins of our universe, are much more eagerly embraced by the media than concepts that claim that solving these mysteries will never happen. It indicates the desire for a 'controllable, engineered' world. A certain complacency seems to have seeped into our minds, as we are shutting ourselves off for things we cannot know and therefore not control. We want to be the lords of our lives, without risks and without fate. Complete control is what we long for. At the same time, this vision upholds the illusion that in the West and other rich parts of the world we can still become richer without serious consequences, that there are no limits to what the Earth can offer us, and above all, that we do not have to share with the rest of the world. The facts suggest, however, that we will have to share if we do not want to immerse the rest of the world in poverty and hunger. In doing so, we ensure not only a fairer distribution, but also more stability, peace and a reduction of refugee crises. Sharing is therefore a necessity. For example, there is not enough water in the world to cater for a Western lifestyle for everyone. There is, on average, 2,300 liters of sustainable water per person per day available in the world; an average Dutch person currently consumes 4,000 liters per day. This lavish water use can predominantly be attributed to our meat and clothing consumption. This involves water from outside the Netherlands and is largely unsustainable as it is usually extracted from groundwater sources that are being depleted. Such a water footprint does not seem compatible with a global vision proclaiming that we do not have to share.

With this book, I will argue that we are facing a very interesting era in which we will have to share our Earth with 10 billion people. However, there will only be room for 10 billion people if we become less self-centered. If we manage to live together with 10 billion people, we must learn to share. All in all, this is an optimistic vision for the future; we should not think in terms of doomsday scenarios, and we should not assume that there is not enough. We should not become cynical and assume that everything revolves around money, greed, power and envy. Our aim should be to seek a higher consciousness within ourselves, as humanity has done for 50,000 years. Less self-centered people need fewer items and are better able to share. This sharing is crucial, because we cannot solve these problems with large-scale technological innovations.

The major threats that the Earth and mankind now encounter are generally known, but because of their scale we no longer know where to start; every individual contribution appears to be futile. It seems that we need a new vision on the role of mankind, in order to set new goals. In this book I propose an overture for such a vision and offer a guide as to what we can do ourselves. In my view, there is no reason why we should not be optimistic about our future.

2

Only when the last tree has died
And the last river has been poisoned
And the last fish has been caught
Will we realize that we cannot eat money

Cree Indian proverb

10 billion people

It is a well-known fact that water and food are the basis of all civilizations, as is nicely summarized in the famous statement: 'there are only seven missed meals between order and chaos'. The depletion of one of these sources has been the end of many past civilizations. What most people don't realize, however, is the fact that our current water and food consumption draws heavily on the Earth. The huge quantities of water and food – and also energy – used for human consumption have a major impact on the environment. We often describe this impact as a 'footprint'.

The two most well-known footprints are the water footprint (measuring our impact on the global water supply) and the carbon footprint (measuring our CO_2 and other greenhouse gas emissions, such as methane). The CO_2 footprint is very much in the spotlight due to the relationship between CO_2 and climate change. According to scientists, we can only bring a maximum of another 1,000 billion tons of CO_2 into the atmosphere, if we are to keep the rise in temperature below two degrees. The two-degree increase is seen as the maximum before the Earth will be struck by all sorts of calamities, such as hurricanes and floods. Currently, humanity emits 40 billion tons per year, so in 25 years' time, we should be well and truly done with emitting CO_2.

This climate change is often depicted as a tremendous apocalyptic scenario. Every day we are fully engaged in various energy saving measures and in generating renewable clean energy. I believe these actions to control CO2 emissions may encourage a false sense of security. By reducing CO2 consumption, it gives the illusion that we are still in control through various technical measures. We build windmills, we produce bio-ethanol and we even think about nuclear energy. In fact, the climate problem has become big business. The Paris climate agreement requires €12,000 billion in investments. Many coal-fired plants must be replaced by other energy suppliers, houses must be isolated, etc. But in the end, even if we are able to reduce CO2 emissions in connection with the climate problem, we still have no solution for the large pile of other environmental problems.

The water crisis

No solution can be found for the shortage of water, and consequently no global agreements are being seriously upheld on this issue, such as the Paris CO_2 agreement. However, according to the World Economic Forum in Davos, it is not climate change that poses the biggest environmental threat, but rather the water crisis. Since 2015, the water crisis has consistently ranked high in the top five of world risks. Even so, it is not widely known that with our water consumption we far exceed the annual amount of fresh water available to mankind. The only way forward is distributing the water supply more fairly and to stop using more water than is available. We cannot 'make' water in large quantities. In other words, we cannot continue to consume as much water as we do now. And because we need water for every aspect of our lives, this will have quite an impact on our prosperity. This is a much bigger threat to our life of ease and luxury than switching to windmills in order to reduce CO_2 emissions, and this may very well be the reason why we hear much less about the water crisis.

In fact, many sustainable energy plans cannot be implemented in the future because there is simply not enough water. Only 3% of all water on Earth is fresh, the majority of which is frozen in polar ice and glaciers; 97% is therefore salt water and unsuitable for agriculture. All our fresh water comes from the oceans through evaporation.

Everywhere in the world lakes, rivers and underground water supplies are drying up, such as the Aral Sea (2000 and 2015, source: NASA)

However, clouds that produce rain above the sea are no use to us. The only clouds that provide are those that produce rain above land and, moreover, only in places where the rain can be used. Unfortunately, a very large part of the rain falls on places where it is of no use to us, like the South Pole. This rain ends up in the oceans through icebergs and is not available for agriculture. Only 20% of the rain can be used by humanity. This rain will partly be absorbed directly by vegetation, and will partly end up in rivers and lakes. Some of it is also penetrates into the ground, feeding the groundwater reserves. In principle, we can use this rain which falls above land in a sustainable way for energy and agriculture. But humanity uses more water than is available. Many lakes have actually already been drained. An example is the Aral Sea in Russia which has dried up due to the irrigation of cotton fields.

There is also fossil water stored in underground reservoirs, water that has been accumulating in the soil for millions of years. These reservoirs are only slowly being replenished, or not at all. This fossil water can therefore be depleted, just like oil. An example is the Aquilla Reservoir that lies under the prairies of several states in America. All their economic and agricultural activities depend on this reservoir, which has already been depleted by one-third. It is a matter of serious concern that, in a mere hundred years, we are now draining these large fossil groundwater reserves, which have been built up during millions of years. This water consumption cannot be regarded as sustainable, since the depletion will be faster than the refilling. There is a severe shortage of water, and as a result about half of all river basin areas are already being over-used and the fossil groundwater is being consumed. Nature also suffers: without fresh water, no nature. Many nature reserves have withered. When there is a drought in Africa, less blue herons return to the Netherlands in the spring. More than half of the people on Earth are already facing water scarcity for at least one month per year.

Often we do not even realize that water is consumed by everything we do in our daily lives. Consumption means that the water has evaporated or is incorporated into the product. Showering does not count, because the water does not evaporate: the shower water returns to the river through the sewer and can be re-used. However, when we spray the garden, we do consume water since it is evaporated by the plants. Our water consumption turns out to be much higher than the two liters of water we drink every day. To produce a kilo (1,000 grams) of grain (bread), about 1,000 liters of water is needed and 1,000 liters of water are also consumed for 100 grams of beef. Food production through agriculture is, therefore, the biggest water user. Producing electricity with oil, coal, gas or nuclear energy also costs a lot of water, because cooling water always has to be evaporated. Cotton, coffee, chocolate, nuts and tea also incorporate high amounts of water.

Arjen Hoekstra, a professor specialized in sustainable water use, has calculated the water footprint for many products. He calculated that the average Dutch person (and the average European for that matter) unconsciously consumes 4,000 liters of water every day. That includes the two liters of water we drink in a day. A Dutch household uses an average of 125 liters of drinking water per day for washing, showering and flushing the toilet. This water, however, returns via the sewer and, as such, is not consumed and therefore does not count. The figure below shows where our invisible water use goes. The majority of our water consumption can be attributed to food, and most of it to animal products, such as meat and dairy. Is this daily average water usage of 4,000 liters by the Dutch really too high? Chinese and Indians use about 3,000 liters per day and Americans 7,000 liters. If we

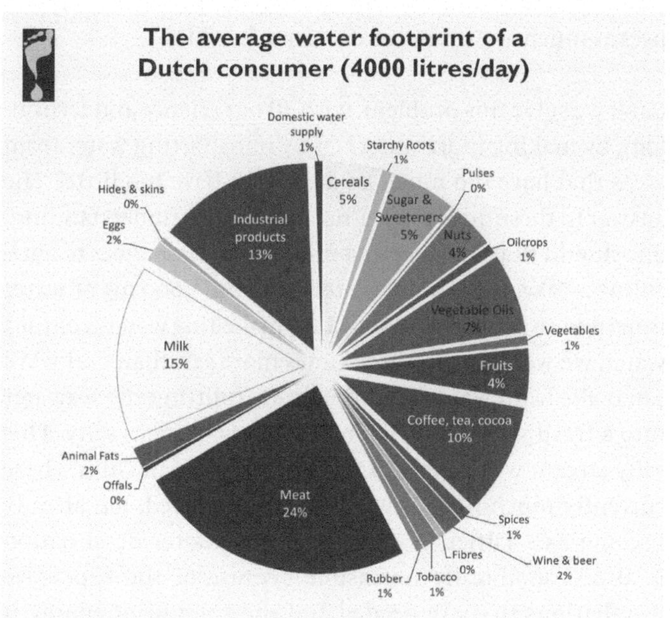

The water consumption of an average Dutch person can be attributed for the most part to food. Water in the household is less than 1% of this consumption (source: Mekonnen & Hoekstra, 2011 – *National Water Footprint Accounts* UNESCO-IHE)

want to stop the global water scarcity, we need to reduce our water consumption to 2,300 liters per day, per world citizen, by the year 2050. In the Netherlands, a lot of grain and soybeans are imported from other parts of the world, thereby drawing heavily on water resources elsewhere. People in poorer areas, who do not eat meat and consume little energy, consume far less water. In countries such as Angola, Guatemala and Nicaragua, for example, water consumption is already under 2,300 liters per person, per day. The question therefore should be: how can we take a step back in our water consumption? Evidently, a scenario in which everyone uses as much as Europeans do, is not feasible.

Can't we solve this problem, with all our science and technology, by making more water? Or by transporting water from areas that have too much, to areas that have too little? The answer to these questions is 'no'. For a better understanding, one should bear in mind that most of the water goes to agriculture. Take, for example, a ton of grain: 1,000 m3 of water is needed to grow this. Should we make this water from sea water, we would produce 35 times more salt than grain. We can make fresh water from the sea by splitting the seawater into a fresh stream and a stream that is twice as salty. This salty stream will flow into the sea. The Persian Gulf, where currently much fresh water is being produced, has already become 25% saltier in recent years. Seawater desalination is also economically unfeasible because of the expensive installations that are needed and the amount of energy it consumes. With a grain price of around 150 euros per ton, for which the farmer must sow, fertilize, harvest, etc., there is little money left over for water in this budget. The water may not cost more than a few cents per m3 (= 1,000 liters). Producing fresh water from seawater costs at least 30 cents per m3, even without counting the cost of transport to the agricultural areas. The grain price would amount to 500 euros per ton, using this type of water, compared to a grain price of about 100 euros today. This is unaffordable for billions of people. By comparison, to put this in perspective: we pay more than 1 euro for one m3 of drinking water in the Netherlands, which is 20 to 50 times as much as a grain farmer can pay for irrigation water. The amount of irrigation water significantly exceeds the amount of drinking water in the world. We drink about 2 liters per day, compared to the 200 liters needed for our cereals and grains, which does not even include the cereals used for our livestock. Drinking water in the Netherlands costs around 0.1 cents per liter and bottled mineral water costs as much as 50 cents per liter, which is 500 times as expensive.

The amount of available water shows us that humanity cannot become infinitely more prosperous. This view is persistently denied by many, because people have the illusion that with innovation and science all problems can be solved. With all our ingenuity and knowledge, however, we are still – and increasingly – confronted with water scarcity, for which we have no solution, whilst water consumption is still on the rise. More and more people are inhabiting the Earth, and thus more food and water is needed. In addition, we are accumulating wealth, and wealthy people use more water than poor people: mainly because rich people eat more meat, but also because rich people consume more energy and buy more items, for the production of which water is also used. Clearly it is a matter of simple arithmetic to demonstrate that in many parts of the world this water scarcity will sharply slow down – or already has slowed down – food production, as well as industrial production.

A water shortage of no less than 40% is expected within 15 years, whereas we already use 15% too much water. In many places, the large fossil water resources are becoming depleted. This has already happened in several areas of the world, for example in Iran in the production of pistachios. In the USA, China and India, major agricultural areas depend on these fossil water sources, which will become empty within the foreseeable future. Already billions of people suffer from some form of water stress and this number will increase exponentially, as the world population grows and becomes wealthier.

El Niño, a natural phenomenon of changing rain patterns, affected an estimated 60 million people who suffered from droughts in 2015/2016. There are concerns that by 2050 there will be about 200 million migrants due to climate change, and the threat of drought. Most migrations due to water problems are confined within the country's borders, as in the case of nomads in Kenya, who move away from their habitats with their cattle because of the drought. This immediately leads to conflicts with agriculture farmers because their harvests are eaten by the cattle.

Growth of the world population

In the 20th century the world population has grown exponentially. At the beginning of 1900, there were 2 billion people on this planet. In the 20th century, this amount increased by no less than 5 billion people: at the end of 2011, we reached the milestone of 7 billion people on Earth. Our number is currently growing with around 70 million people a year; it is anticipated that in the 21st century another 3 billion people will be populating our planet. Evidently, the growth in world population is decreasing. If every woman had 2.1 children, the world population would become stable. Large areas on Earth, such as the EU, Russia, China and Japan, remain under this average. In South Asia and Africa, the birth rate is still much higher. In these areas the population continues to grow strongly (by 2% per year). In sum, however, total growth is slowing down. It is predicted that we will reach a stable number of approximately 10 billion people by 2100, and that after 2100 this growth will stop. The population is estimated to only keep growing in Africa, whereas in Asia and Europe the population will decrease by 0.1% to 0.2%.

With such growth numbers, the theory of Malthus (1798) is often resurrected; he predicted that the population would grow faster than food production, causing a major famine. I think he's wrong. In my view, the world population has always followed mankind's ability to make food, on top of – of course – our ability to live longer through more hygiene and better food. In my view, it is a tremendous human achievement and an icon of cooperation, insight and ingenuity of mankind that so many people can now live on Earth. Never before has humanity been able to achieve this. Not only innovation has made this possible, but also the diminished egocentricity of humanity in the past centuries, because without cooperation it would not have been possible. Various scenarios show that humanity can procure enough food for 10 billion people, provided that we share the available water fairly and everyone consumes less than 2,300 liters of water per day.

Growth of wealth

Not only the population is growing, but our wealth is as well. According to Thomas Piketty, until the year 1700, the rise in prosperity was equal to the growth of the world population, so on average everyone remained more or less equally rich. Between 1700 and 1820, per capita production –and therefore welfare – increased by 0.1%. From there on, things went increasingly faster. Between 1820 and 1913, production increased by 0.9%, and from 1913 to 2012 by 1.6%. These growth figures may seem low, but by repeating this every year, we have become almost 5 times as rich in the last hundred years, per person.

Combined with the growth of the world population, production has increased by 3% in the last century (1913-2012). Converted, this means that the production of humanity has become 25 times higher within hundred years, whereas in the 100 years before (1820-1913), production had already grown 4.5 times. It is good to realize that this growth in production of the world population per capita is just as important a factor for the demand for water as the growth of the population itself. Had everyone remained as rich as in 1900, we would now use 5 times less water.

Moreover, production and prosperity are, unfortunately, not evenly distributed among the world population. People in Europe and North America earn between 2,000 and

3,000 euros on average a month, whereas people in Africa and Asia have to make do with 200-500 euros per month. In Latin America, the monthly earnings are similar to the global average of 780 euros per month. Given the shortage of water and the biodiversity crisis, the question is how much growth in prosperity the world can bear. If everyone wants to become just as rich as we are in the West, earning 2,500 euros, the capacity of the Earth will be overburdened. Theoretically, there may be room for 10 billion people, but not for 10 billion 'rich' people.

Naturally, the majority believe that the people living in appalling poverty must become wealthier. Currently, however, it is in fact the rich countries who feel that they are in dire straits, and want to become wealthier. The richest regions in the world, the EU, the USA and other OECD countries, still strive for growth, preferably by more than 1.5% per year. But if our economy grows by 1.5% for 30 years, while our population remains stable, everyone gets wealthier by 50%. Which is exactly what has happened in the past 60 years. The book *Gouden Jaren* (Golden Years) by Annegreet van Bergen describes what the increase in prosperity has meant for the Netherlands since 1950. As a man in his fifties, I myself have experienced by and large this increase in wealth. The availability of all kinds of fresh fruit and vegetables throughout the year, regardless of the season, is an example of this. Another is the luxury hospitals that we now have. Louis XIV (the richest man in the 17th century) would have been jealous at our modern houses, with heating, lighting and running water. Our modern mobile phones are already more powerful than the first personal computers. TV has dozens of channels, 24 hours a day, whereas in the past we only had two channels with, at the most, 10 hours of broadcasting time per day. Moreover, almost everyone has a car, which are much more luxurious than the first models.

All this wealth, however, also leads to waste: we throw away as much as 30% of our food. What will our life look like if we grow 50% wealthier? Are we going to throw away 70% of our food? Previously, we would buy and eat an entire chicken, but now we only eat the breast and legs at most; the rest of the chicken is lost for human consumption. That means we have to breed more chickens. What will more wealth mean in the future? Are we all going to buy robots? Are we all going to automate our homes? Is all this really worth it to risk an unprecedented environmental crisis?

Finally, animal welfare has not improved in the last 50 years. Apparently we can no longer afford better conditions for chickens, despite the enormous increase in our wealth. When we were poor, the chickens were allowed to roam freely on farms. Now that we are rich, 150 million chickens per year are bred and slaughtered industrially in the Netherlands alone. What's wrong with this picture?

One wonders what it means for the lives of people in the rest of the world if we accumulate another 50% more wealth. In this scenario, I see at least three major disadvantages.

- Firstly, such wealth also means poverty in our daily lives. All jobs in which it is not possible to increase productivity will become unaffordable. Productivity is seen as the amount of product an employee can make per day. In industry and agriculture, with more and more automation, this increase has been considerable, but with other jobs, such as primary school teachers, the productivity growth is virtually zero. There is a limit to the amount of children one teacher can take care of. Other examples are postmen, cleaners, nurses, help for the elderly, etc. These professions cannot be made much more efficient. This is already clearly visible in the Netherlands. For years, the salaries in these professions has been below the increase in welfare. Swimming lessons are withdrawn from the school curricula; although the Netherlands have never been so rich, these are now too expensive for schools. Post offices have disappeared. In the healthcare sector, everything creaks under the weight of personnel shortages. Nowadays many have flexible contracts, although research has shown that these contracts often make people insecure and unhappy. Our wealth will therefore actually turn out to be poverty if we no longer have help for the elderly, nurses or swimming lessons.

- Secondly, our prosperity growth also comes at a cost. A few examples: clearing soil pollution counts as economic growth, but it does not add to our prosperity. Due to our enormous productivity, there are increasingly more people who cannot keep up and will be left behind. Despite the fact that citizens in rich countries score very high when asked whether they are happy, they score low on the question whether they experience a sense of purpose. The richest and happiest countries use enormous amounts of antidepressants and have to deal with a high rate of suicide. In the poorest countries of the world, where happiness scores lower but meaning scores higher, there are far fewer suicides. Further, the productivity gains of automation by computers and robots are also under pressure due to problems with hackers. The most modern container terminal in Rotterdam, the largest sea harbor of Europe, was shut down for a week in June 2017 by a hack. The problem was even more serious, as nobody could operate this terminal manually. A month earlier, hospitals in England were shut down for the same reason. Investments in cybersecurity will have to be increased considerably, thus offsetting part of the profits through automation. Costs of the increase in welfare are estimated at 18%. We can save ourselves these costs if we were to choose to be slightly poorer, and seek meaning rather than only wanting to accumulate wealth. Also, any further growth in wealth only seems to benefit the rich and thus increase inequality. This appears to be a kind of pattern in capitalist societies, where land, labor and money are traded through markets, as in our western society. Bas van Bavel has described this pattern. It starts with an open society in which everyone has opportunities for growth in prosperity. Soon after, inequality increases sharply and the rich obtain political power. As a result, economic dynamism disappears and a downturn occurs, with increasing coercion. Compulsory military service is abolished, mercenary armies appear and civil rights decline. He found this pattern in the capitalist societies of Baghdad (500-1100), Italy (1000-1500) and the Netherlands (1100-1800).

- Thirdly, the Earth cannot sustain this wealth, either with 7 billion, or with 10 billion people. Our current water consumption will lead to a major shortage in the future. When the water is depleted in large areas of the Earth, billions of people will start to move. That can only end badly for all of our planet's inhabitants. If the West continues to grow wealthier, it will also be an example for the billions of poor people. They also want more, or will become frustrated by the large differences. Rather, if we want to reduce our water footprint on the Earth and distribute the water fairly, the West will have to seek some degree of austerity. This means that we should only pursue growth if it requires less water and results in a smaller footprint: this will be the condition for further wealth.

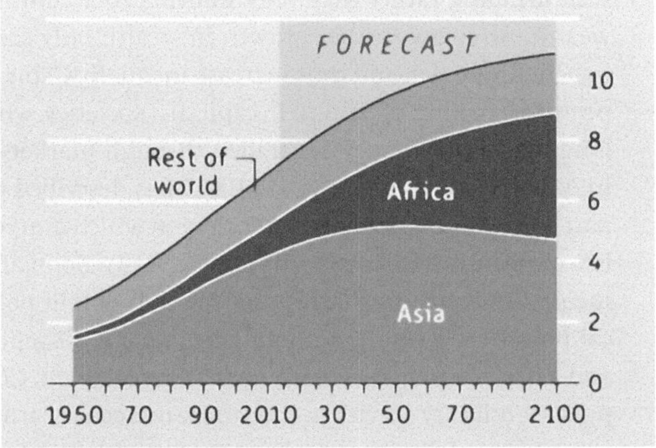

The world population will grow to around 10 billion in 2100. The population will continue to grow strongly in Africa and Asia (source: UN)

An important question related to this topic is: don't we need economic growth to solve the poverty in the rich countries, because even in the richest countries more than 10% of the population has to live in poverty? In my opinion it is not the lack of wealth that causes this poverty, considering we are talking about the richest countries in the world. Becoming even richer will not solve the poverty in these countries, as it is clear by now that the increase in wealth doesn't reach the poor and will mainly go to the rich side of society. As mentioned before, because an increasing amount of people cannot keep up with the growth in productivity, poverty will only increase with growing wealth.

The will to share?

Many people are confident that we can solve all environmental problems with technology and innovation, and that we can simply become richer without problems. Yet, in my view this is a utopia, since the water problem is not the only problem. There are many environmental and social problems that only get bigger, despite – or thanks to – our wealth. In addition to the climate crisis and the water crisis, the World Economic Forum also mentions geopolitical tensions, the involuntary migration flows, and environmental problems. On top of that we have the pollution of coastal seas, rivers and the Plastic Soup in the oceans. There are deadly air-borne fine particles in cities due to intensive traffic, and there is an imminent shortage of all kinds of essential raw materials, such as zinc for agriculture. The loss of fertile soil is also a major threat. Only 3% of the land surface on Earth is fertile and suitable for agriculture. Vast areas of fertile land are lost every year due to over-fertilization and salinization. If there are still optimists who think that these problems can be solved, they should consider the story of the biodiversity crisis. Mankind is overfishing rivers and seas, is cutting the primeval forests and jungles, switching to industrial agriculture with monocultures and all sorts of pesticides, and pollutes much of their natural environment. Several studies show that 75% of all insects have disappeared in European countries like The Netherlands and Germany. In my opinion, all of these social and environmental problems are caused by the same fundamental attitude: the reluctance to share. I have chosen to concentrate on water shortage, since this problem will best convey the message that we need to share, which means living a more sober lifestyle, simply because there is no other solution. We can try to resuscitate forests to collect more water and retain it in the soil, we can grow crops that use less water, but ultimately we will have to share in order to solve this immense and complex problem.

My opinion, however, may be controversial. We continue to focus heavily on the economic growth of the richest countries and we cling to innovations to save us. It is often claimed that more money is needed to save the jungles and the environment. That seems strange and not compatible with the fact that in the last 100 years, in which humanity has become 25 times as rich, all problems have grown in size. Wealth and waste seem to be the cause of the problems rather than the solution.

There seems to be an egocentric, narcissistic attitude in the West (and in other prospering regions in the world). People want to focus primarily on their own comfort and luxury and do not care much about what that does to others. An example is the widening of the E25 (the motorway between Utrecht and Amsterdam in The Netherlands). The residents have agreed to this widening and thus with many more cars driving past their homes, provided that the maximum speed is limited to 100 km/hour. After the opening of the road there were loud protests from the motorists claiming the road was too alluring for a maximum speed of only 100 km/hour. Now the maximum speed is 130 km/hour at night, without any thought for the inconvenience this may cause for the local residents.

Could such narcissistic and indifferent behavior be caused by the idea that we were created by pure chance in an immense universe, whose train of thought is prevalent in Western cultures and has nestled in our minds? In other words: if life does not make sense and there is no afterlife, what does it matter how we live? In that case, we might as well live selfishly. How do we rid ourselves of the narcissistic pursuit of wealth alone, which leads to an environmental crisis that will affect the poorest in particular? Wouldn't it be great if we realize that more consciousness would be a better goal and more rewarding than trying to accumulate more wealth, since obviously getting richer is also the cause of several fundamental problems.

3

*The most beautiful and deepest experience a man
can have is the sense of the mysterious. It is the
underlying principle of religion as well as of all
serious endeavour in art and science. He who never
had this experience seems to me, if not dead, then
at least blind. To sense that behind anything that
can be experienced there is a something that our
minds cannot grasp, whose beauty and sublimity
reaches us only indirectly: this is religiousness.
In this sense I am religious. To me it suffices to
wonder at these secrets and to attempt humbly
to grasp with my mind a mere image of the lofty
structure of all there is.*

Credo Albert Einstein

Science:
we cannot know everything

How many books can a person read in his life? That varies per person, probably ranging from hundreds to thousands. Imagine being in a library with 100,000 books. That is just a tiny fraction of the number of books there are in the world. With so many books, we realize that it is impossible to know everything that is written down. And with the pace at which new scientific articles appear – millions per year – it is also impossible to keep track of what is added to the pile of scientific knowledge every year. In the face of all this knowledge, we become modest and realize how much we cannot know. We become even more modest when we realize how much has not yet been written down. There are an infinite number of subjects that we, as mankind, have not yet understood or haven't even become aware of.

We can also see that there are things that are too complex for our intelligence and are therefore 'unknowable'. So it seems pretty wise to admit the fact that we – as mankind, but certainly as an individual – actually are still ignorant on most matters. The same applies just as much to science. We know for a fact that one plus one makes two, but that is not an answer to an important question. Generally speaking, the more certain science is about things, the less important these tend to be for the current major societal themes. Suppose we want to investigate whether an apple is healthy. Health is an infinitely complex phenomenon that is influenced by our social environment, pollution in our environment, lifestyle, our genes, etc. An apple is also infinitely complex in composition. The difference between an apple grown according to the organic farming method and an 'ordinary' apple already leads to many unresolved questions. Do we look for substances in an apple, such as vitamins that we recognize as healthy? Or are we going to ask a sample group of a thousand people to eat an apple every day, and another group the same size to not eat an apple every day? In order to see whether, after a year, they are equally healthy? Or should we only look after ten years? Unfortunately, such an experiment cannot be sustained. Perhaps the effects can only be seen in the next generation. Yet there are many scientific publications and research on the subject of whether an apple is healthy. On the subject of health there are so many publications that some researchers only analyze what has been published before. If a majority of the earlier publications declare apples to be healthy, then such a scientist considers this to be true. There is also the problem that many studies have not been carried out properly and other scientists do not trust the conclusions of previous studies. A major problem in science these days is the increasingly large fraction of experiments that are not reproducible.

We also have to be aware that a scientist at a university or a public knowledge institution is supposed to publish scientific papers; the more amazing, the better for their careers. The societal and economic consequences of such publications rarely reach the scientist. If, for example, a scientist claims that apples are not healthy and people subsequently eat fewer apples, causing apple growers to go bankrupt, it will not have any consequences for the scientist's career, even if other scientists prove that apples are healthy later on. So whereas society is vulnerable to the outcomes of science, this is not – or hardly ever – the case the other way around.

In all this, we assume that scientists are objective and have no interest in the core outcome of their studies. But if the scientific study, for example, has been paid for by the apple growers, can the results still be trusted? Especially if the conclusion is to eat more apples? When determining whether medicines are safe, we see that things go wrong every now and then, as with the drug DES which led to birth defects, or the more recent anti-inflammatory drug Vioxx that led to heart problems. Clearly, the scientific studies in these cases were influenced by the money from the big pharmaceutical companies. Were these scientists able to determine the safety of these medicines with the current state of scientific methods? And were they trustworthy, or did they have an interest in the outcome of their research? A lot of research at universities focuses on topics supported by industry, who are mainly interested in innovations that they can earn money with. This is also the case when these innovations are not necessarily the best solution for a problem, such as the claim that an expensive pill is better than a cheap apple.

Even if scientists are completely honest, they may have a certain line of thought that partly determines the results of the research. Where else would the research questions come from? The assumptions that determine the basis of this line of thought are often not even clear to the researcher himself. Although these reflections put the value of science into perspective, we increasingly expect that science will provide answers to the big questions and, together with technology, will provide the solution for the biodiversity and water crises.

Everything is connected in an infinite chain of cause and effect. Sometimes we can see bits of the chain, for example if we describe a food web in nature. Meat eaters eat plant eaters or insect eaters. In turn, insects eat plants. Plants, for their part, eat the feces of the animals, which first have to be digested by bacteria and fungi. There are hundreds of species of meat eaters, thousands of insects and millions of species of bacteria. With this chain we only look at who eats who. We are not yet looking at the influences of the sun and the moon on ocean currents that affect habitats. Or the effect of minerals on all kinds of plants, etcetera. In the 1970s, the number of birds of prey in the Netherlands suddenly decreased. It turned out that the shells of their eggs were too thin. This was attributed to toxic industrial waste that was dumped into the sea by the industry. Through all sorts of animals, this poison ended up in the birds of prey. Was this scientifically foreseeable?

The big question now is whether or not we believe that science is capable of fully describing the infinite chain of cause and effect, in order for us to know what the consequences will be from all our actions or plans. For this, all links must be reduced to much more simple descriptions than reality. Science regularly makes it seem as if this infinite chain had already been charted. I do not believe that science can trace the infinite chain of cause and effect. As a wise statement goes: the chain of cause and effect is not more complex than we think, but more complex than we can think. Take, for example, the great mysteries of life. There are mysteries that can't be solved by science, despite claims by scientists that eventually everything can be explained. I very much doubt that science will become omniscient and I therefore feel that science should become more modest in its statements. Let us examine this further on the basis of two phenomena, namely the origin of matter and the origin of life. In doing so, I make a distinction between phenomena that (in my opinion) will never be solved and will remain a mystery, and phenomena that cannot be solved at this moment, but possibly, however inconceivable currently, will be solved one day.

Origin of matter

And all of a sudden arose, from scratch, all matter that makes up the entire universe. The scientific hypothesis is that at some point, 13 to 14 billion years ago, out of nowhere there was the Big Bang. At that moment, something suddenly arose from nothing. And that something contained all matter of our entire universe, first in the form of hydrogen atoms and later in the form of all elements, molecules, stars and planets. It took no less than 10 billion years before the Earth was here. There are billions of stars in the universe. Not only the matter to make all of these stars was present, but also the information to shape our current universe, for example all the laws of nature and fundamental parameters of nature, such as the mass of a proton. Had these parameters been only slightly different, the universe – as we know it – would not have come into being. It would have collapsed, or there would never have been stars and planets. We also assume that at that point time started. Since time and space are connected, we cannot look beyond the universe, and that means that we cannot look back beyond the time when the universe was created. We do know, however, that the universe is getting bigger, but we do not know in what space around the universe this happens.

It is well recognized that it is impossible to get information about the time before and during the Big Bang and that the question where matter comes from cannot be answered. We can classify this as a mystery. Nonetheless, some scientists persist in their belief that they can eventually come up with a logical explanation. Be that as it may, this cannot be based on science, whereas they do – above all – believe in a future scientific explanation. Although, several assumptions have been put forward. One answer is, for example, string theory, which claims that there are multiple universes and that it is therefore no coincidence that our universe has exactly the right constants of nature needed for developing the right elements, and for stars and planets to develop, providing all conditions that ultimately make life possible. However, for this theory to work no less than 10500 universes are needed. Like I said, this is merely a belief.

Another point is that the reputation of science has been acquired through the *description* of the laws of nature, whereby, as a rule, the novel mathematical formula that describes a natural phenomenon most compellingly and simply, replaces the previous formula. A recent example is Erik Verlinde's theory, which explains gravity with density of information, instead of Newton's theory, which did so with mass. If Verlinde can convince his academic peers, his description will take over. Scientists' standing can also be improved further by predicting something on the basis of a theory that has not yet been measured or observed. A well-known example is Einstein. He predicted that the deflection of light by the sun would be between 1.5 and 2 degrees. When it became possible to measure this, it proved to be 1.75 degrees, right within that range. It was all over the newspapers. The discovery of the planet Neptune also has such a history. From calculations by the French mathematician Urbain Le Verrier on the basis of the orbital path of Uranus, Neptune could be localized. There is also a famous example in chemistry. Mendeleev developed the periodic table, in which he arranged all elements in a logical order. In this system he left spaces for elements that were not yet identified. In this way he has predicted dozens of elements that have only been discovered later.

That is all well and good, but in my opinion, the reputation of science is unduly increasing as we often read in the newspapers that science has found the cause of something (e.g. gravity), whereas it has actually only found a way to *describe* the phenomenon. Take, for example, a prediction from quantum theory that can only now be demonstrated: we can now show that particles are bi-local. Two particles at a larger (to infinite) distance continue to show the same characteristics, no matter how far apart they are. We can prove it, we can describe it and we even think we can make a computer with it (the quantum computer), but it is impossible to explain. It is totally incomprehensible for us. We tend to think in 'cause' and 'effect', in which 'cause' always has to come first. This is no longer possible with bi-locality. These particles change simultaneously (at least faster than light). This makes us wonder whether thinking in cause and effect is, after all, the right approach to understanding reality.

The same applies to cosmological models, in which we, for the sake of convenience, prefer to assume that 96% of the universe consists of dark energy and dark matter. As yet, however, we have no idea how to measure or demonstrate this. According to this model, we are only aware of 4% of all matter. An average newspaper reader would hardly believe that our smart physicists can only talk about this '4% of everything'.

Last but not least, there is also the intriguing evolution of the elements. Everything originated from hydrogen and all elements consist of protons, electrons and neutrons. But we cannot explain why the combination of these particles produces chlorine, which is green and highly corrosive, and that a slightly different combination of these neutrons produces sulfur, which is yellow. Where did all of this molecule information stem from? There is still hope that we can gain insights from the particles that make up the atoms, such as the quarks and glucons. However, this does not yet explain the behavior of the atoms. We still have to look further. The Higgs particle was discovered most recently, thanks to the Large Hadron Collider of the research institute CERN near Geneva. This is a particle accelerator of no less than 27 km in length. The experiment has cost more than 10 billion euros. Have all questions been answered with this experiment? The answer is 'no'. As is often the case with new answers, new questions arise as well, such as: does the Higgs particle consist of even smaller particles, which might describe the properties even further? For these experiments, larger particle accelerators are needed, which cost an astronomical amount of money and require unimaginably large amounts of energy. But these smaller particles will also be made up of something. Is it conceivable that we will ever find the smallest particles and can we still pay for this type of experiment?

Origin of life

The second phenomenon that I want to discuss is the origin of life. Every time a fossil is found that changes the existing theory, we often hear the comment that we are one step closer to unraveling the origin of life. I believe that is incorrect. Life has, as far as we know, only emerged once, and this happened about 3.5 billion years ago in Australia. Stikker interviewed several scientists on this topic. How life emerged remains a mystery. Some scientists believe that this can occur spontaneously, as a result of the theory of the primordial soup in which organic structures emerge spontaneously, under extreme circumstances, and become alive. If something like this can happen out of the blue, it should have happened at more locations and times, but we cannot find any proof of that. Even with all our current technical means we cannot generate life, even if we would try this, for example, with a dead bacterium that had all of the complex molecules in the right place. We can only apply our genetic methods to living organisms. We cannot give life to anything that is not living. With the emergence of life, therefore, a lot of information has had to be generated in one go. The already very complicated codes from the simplest and oldest forms of life are still to be found in all life forms.

Darwin's theory is often confused with the origin of life. Darwin's survival of the fittest is about evolution – how single-celled organisms, plants and animals became increasingly complex – and not about the origin of life. There is hardly any way to investigate the origin of life. For this, there are no particle accelerators which offer a way forward. A lot of research, however, is being carried out into DNA molecules and genetics. The speed with which we can read DNA molecules has increased exponentially, whereas the costs have spectacularly decreased. The Human Genome Project has mapped the human DNA. The hypothesis was that everything that is alive can be traced back to the DNA that is present in that life. So knowing your DNA is knowing you. However, this does not appear to be the case. In the DNA of humans and chimpanzees there is only a 2% difference, whereas the variations in appearance and capacities are considerable. The information contained in the DNA appears to be used mainly for supplying chemical products and processes in living beings. Where the information comes from that causes the differences between all plants, animals and people remains unclear.

In conclusion, the origin of all matter and the origin of life, including constants of nature and the code for the first DNA, remains unattainable for scientific research. Suggesting that this is indeed attainable must, at present, be based on a belief in science. A belief somewhat comparable to the belief in a creator. That we can unravel the structure of matter and the associated properties or the information carrier of all properties of living beings, may very well be possible, but at the moment it seems utterly inconceivable.

Limits to controllability

As a child, born in the sixties, I have always been under the delusion that all useful and good things were invented only after World War II. During my childhood, one technology after another entered our home, such as the color TV, plastics and the tape recorder. Before that, in my childlike eyes life had to be miserable. In a way that makes sense, because everything new gets attention on TV and in the newspapers, and everything old does not. Now I know that electricity has been around for a long time and so have trains; and that many of the physical and chemical breakthroughs were pre-war, such as the production of nitrogen fertilizers from the air, and that virtually all of our everyday items – such as pots, pans, cutlery, shoes and clothes – have been the same for thousands of years, only now perhaps made from different materials. Apparently, I am not alone in my craving for the modern, for the new. Everyone seems to be affected, with the result that we not only think that the modern is better than the old, but also that science and technology can solve all problems. This can be misleading.

Technological possibilities have given us the idea that fate has come under our control. However, we still cannot predict earthquakes and many other disasters. By making statements on these events, often reassuringly, many people have already gotten into trouble. Take, for example, the lawsuit against seven Italian scientists who had predicted just before an earthquake that there was no immediate danger of an earthquake. They were convicted of manslaughter for insufficiently warning the residents of the town of L'Aquila in 2009 of an all-devastating earthquake. The seven scientists, all members of the Large Risk Committee, were each sentenced to six years in prison. This condemnation was unprecedented and protests came from all over the world. Eventually they were acquitted on appeal. Another example illustrates this point further: our economic-financial scientists are still unable to predict an economic crisis. Reassuring statements made by renowned economists at the beginning of 2008 have put many people on the wrong track, leaving them unprepared for the economic crisis. Moreover, we do not yet have a cure for the majority of all diseases. Yet many people feel that they can be 'repaired' for almost anything in a hospital. In that case they are in for quite a disappointment.

New technologies can also prove to be damaging. Plastic wrappings and clothes, like fleece sweaters lead to the plastic soup in the oceans, where enormous accumulations of plastic waste affect marine life. The more complex the problem, the harder it becomes to predict the harmful effects of our actions. Our health and natural environment are two such complex issues. For example, Nassim Nicholas Taleb describes the harmful effects of the medical sector on health in his book *Antifragile: Things that Gain from Disorder*. Any medical intervention may also have negative effects, and these outcomes are not linear. Minor medical procedures can have serious and even fatal consequences. In the Netherlands, people are much more likely to die from medical errors than from traffic accidents. This is why Taleb advises us to only seek medical attention in the event of life threatening diseases and accidents. An important result of the healthcare industry is that we will all live a few years longer, but the price for this is that we all become slightly less healthy. Given the number of 7 million people suffering from a chronic disease out of a total of 17 million Dutch people, he seems to have a point. In the Netherlands alone 4 million people are taking anti-hypertensive medication, 1.5 million are using cholesterol-lowering medicine, 1 million suffer from chronic lung diseases and 1 million people have Type 2 diabetes. Also the healthy life expectancy, which is the time after our birth that we are not sick, is decreasing (currently below fifty years in the Netherlands), while life expectancy continues to increase (currently above eighty years). So in this respect it is not only positive what science can bring us. The sacrifice of less health during life receives very little attention compared to all kinds of medical breakthroughs that are widely reported and commented on in the media. Taleb defines the concept of antifragile, as the opposite of fragile (vulnerable). All things and habits that survive for a long time are antifragile, since they have survived all sorts of shocks and changes. The older something gets, the more worthy of respect it stands. Taleb suggests heuristics (traditional common sense) above science. According to him, for example, we should take a walk every day, because that is what people have been doing since our creation, until about three generations ago in the West when we almost stopped walking altogether, at first because of the use of horses and later the use of cars. In his opinion this cannot be healthy. And when using a bit of practicality, this makes perfect sense.

Nature, which has existed for billions of years, does not have to prove itself. This is in opposition to all of the unnatural things we want with regard to ecology, agriculture, food, health and healthcare. The absence of harmful effects in the short term is, therefore, no guarantee for the long term. The greater the effect of a new technology, the more careful we ought to be with it. A nuclear disaster, for example, wipes out all the profits that have ever been earned with this technology in one go; the costs of Chernobyl and Fukushima are running into the billions. The effects of unnatural developments, such as genetically engineered agriculture, should also be followed for a very long time before being applied on a large scale. On that account, the consequences of the Green Revolution are gradually becoming apparent. Modern cereals are dependent on nitrogen fertilizer which, in turn, has to be made with fossil fuels and affects soil biodiversity. A consequence is that the plants are more susceptible to diseases and they can absorb fewer substances from the soil, making them less nutritious. These cereals also require much more water. Dan Barber, a chef from New York known for his sustainable cuisine, describes in his book how in India the wheat varieties developed during the Green Revolution are clearly connected to the water crisis. They yield 40% more wheat per hectare, but cost 300% more water. The switch to modern cereals has also resulted in less oats and barley on the menu. This was apparently a self-reinforcing effect: the more grain, the less oats and barley. But combining oats, barley and wheat is healthier than just eating wheat. Also the cultivation of a combination of oats, barley and wheat is better for the environment and the soil will produce a higher nutrient yield. The absence of harmful effects in the short term is no guarantee for the prevention of trouble in the long term. The Mennonites in the USA claim, as rule of thumb, that the education of a child starts 100 years before birth. This means that the culture, the environment and the surroundings in which a child is born have been shaped for generations before. Such a precautionary principle is something our modern world seems to have abandoned.

The optimism that science and technology are going to solve everything should therefore be put in perspective, because so far it is open to a good deal of debate. With the many scientific and technological solutions, the decline in biodiversity has not stopped and our water supply has only decreased. Modern large-scale agriculture has aggravated these problems rather than reduced them. With big data and quantum computers we cannot conjure up water either. It appears to be vitally important that we do not lose sight of the infinite complexity of nature when trying to deduce it into bite-sized chunks that we can understand.

In science and technology bad luck cannot be ruled out either. Fate can strike anywhere, anytime. A traffic accident, a terrorist attack, a pilot who wants to commit mass suicide, a deadly disease, a tile blowing from the roof: we do all we can to try to get everything under control, but we will never succeed. Therefore, it is for that very reason that we should respect fate. It will mean less fear of what can happen to us and a little more gratitude for everything that goes well in our lives.

Place of science: I, We and It

Many philosophers see the world as more than just the objective truth of science. In this respect, they often distinguish three segments. Ken Wilber, for example, distinguishes between consciousness (I), culture (We) and science (It). In doing so, reality can be better described with the mix of the three, than with the It of science alone, as is the tendency of our contemporary society. Many philosophers have used this classification before, from Plato (before Christ) to Habermas (in the 20th century).

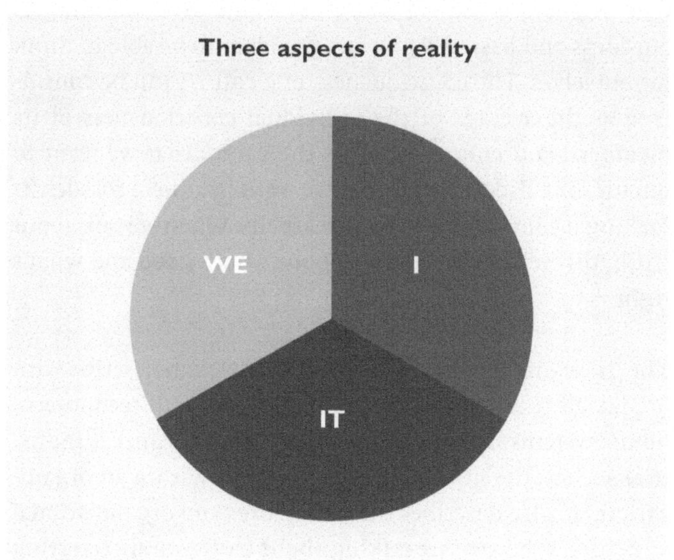

Reality can be described in three segments: the objective It, the subjective I and our common values, We (Ken Wilber)

The 'I' segment is about our inner self. We can only talk about our inner self when we use first-person 'I' language. This is about our consciousness, subjectivity and tastes. When we talk about truth, we, in fact, talk about a subjective truth, and sincerity. For example, someone can lie deliberately, but someone can also lie unconsciously, due to inner experiences that are repressed or distorted. Sincerity requires us not only to be consciously honest, but also to search for the repressed truths within ourselves. Growth of consciousness also falls within this segment.

The 'We' segment is about our culture, the spiritual space that we share as individuals. Our culture gives meaning to our ideas and basically gives us the ideas to be able to think for ourselves. The consciousness of a culture can be considered as the average of the individual consciousness of its members. Our culture gives us the values that we want to pursue, but also the rules that we want to adhere to, such as making it compulsory to wear seatbelts. When we talk about truth, this segment is mainly about what's good and what's right.

The 'It' segment comprises everything we can describe with 'it', i.e. all technology and natural sciences, all techno-economic systems such as capitalism or communism, the industrial society, the information society, etc. Within an organization, 'It' also describes the procedures and organizational structure. When we are talking about truth, we are referring to the factual truth, the truth that is objectively verifiable. In our Western society the 'It' segment is the most dominant by far. If someone is depressed, the 'It' segment will seek a scientific solution and strive for an explanation on a material level, for example with brain scans, and subsequently will supply pills, or a physical treatment using electrodes or magnets. Whereas the 'I' segment approach would include psychotherapy and trying to treat the depression by counseling and interviewing, asking the patient about his or her perspective, without using physical or chemical measuring devices. In doing so, we might find, for example, loneliness to be the cause of the depression. From the 'We' segment, the solution could be motivated by the way in which people interact in the neighborhood the patient lives in; by creating a culture in which people socialize and take care of each other, and thus reducing loneliness among the residents.

Scientific visions have gained great admiration and respect in our current society, often disregarding the 'I' and 'We' perspectives. They influence the ideas we have. We cannot live without ideas; they determine our thoughts and actions. Our ideas enable us to do one thing and refrain from doing others. Ideas from science that we have been hearing from childhood and that influence our thinking and acting, are for example mentioned by Schumacher:

- The idea of competition and natural selection, where the strongest, or the best adapted, continue to exist. From this point of view, competition is good; wars can strengthen society and the weak will have to fend for themselves.

- The notion of positivism, which states that all valuable knowledge can only be derived from natural sciences.

This idea denies any possibility of knowledge about the meaning and purpose of life. From this perspective, man is but a fluke, without real purpose on Earth.

- The tenets of Marx, which label all higher manifestations of human life, such as religion, philosophy and art as fabrications of our brains, serving nothing but the economic interests of others; or Freud's ideas, which alternately reduce these higher manifestations to the dark stirrings of our subconscious.

The influence of these ideas in our modern world is considerable. Yet they are not the result of factual research. They cannot be verified. They are a giant leap in the imagination, with a strong claim to general validity. These ideas are fixed in our heads. All of these ideas have in common that they concern matters of a higher order, but in reality are nothing but a manifestation of the lower. In fact, the difference between higher and lower is denied altogether. We increasingly tend to interpret the world with these ideas, which blinds us to the differences in levels of consciousness or meaning. As a result, the concept of 'consciousness' has become invisible in our world, and insight into consciousness plays too small a role in our current efforts in solving the major environmental issues, such as the water crisis. In chapters 5 and 6 of this book, I will focus on this invisible consciousness in order to fill the serious gap in our thinking caused by these deep-seated ideas handed to us by science.

4

'Can you prove that transfats are harmful?' is a question that used to come up frequently in the past, implying that, its harmful nature had to be proven – which obviously could not be done because it took decades before the damage was revealed. Such questions are often posed by intelligent people.

If the current inhabitants of Mother Earth want to do something that goes against nature, then it is up to them to come up with proof that it is not harmful.

Nassim Nicholas Taleb

Conscious innovation strategy

We observe that as humanity we are slowly but surely becoming surrounded by all sorts of social and environmental problems. Despite all of our environmental technologies and product improvements, our ecological impact on the Earth has not diminished; rather it has kept on rising.

From time to time, the impact has remained the same. This, however, can be explained by a production decrease due to an economic crisis. Every time we create a cleaner product, we almost immediately lose the environmental benefit derived from this innovation. Suppose we produce 90% cleaner cars, but the number of cars increases tenfold, the environmental pollution will remain the same, as we can see in many cities. If we make a car 10% more efficient, but subsequently add some luxury features and more weight, that environmental benefit will soon evaporate. The advantages our energy efficient refrigerators and LED lights give have already been fully consumed by our smartphones. Many of us barely realize how much energy is needed for our smartphones. For all of the photos and videos, huge data servers are needed to store them in the cloud. Every time we download a photo or a video, it has to be retrieved from these large data servers. The data centers in the Netherlands alone (consumption of 2,000MW in 2017) need two coal-fired power plants, or all nine 700MW wind farms that are planned in the Dutch North Sea. Moreover, only 30% of the wind turbine capacity is actually turned into electricity, as the wind is not constantly available. In other words, all new green electricity is already consumed by something that hardly existed 10 years ago. On top of that, the GSM network is constantly active, which also uses up a lot of energy. The higher the speed of the network (from 3G to 4G), the more energy is needed. A smartphone consumes nearly as much energy as an average household through data centers and transmission masts. We all have high expectations of digital coins such as bitcoin, until we realize that one bitcoin transaction costs about 15 kWh. This is equivalent to the energy needed to lift fifteen blocks of 1 m3 of concrete sixty meters high. Even so, many people think and hope that we can solve our environmental problems with science, technology and innovation, and at the same time live increasingly wealthy and luxurious lives, whereas the facts indicate that we are overexploiting fish stocks in our seas, cutting down our jungles, destroying the fertile soil and depleting our water resources. It seems more likely that the opposite will happen, that the environmental problems are going to slow down our economic growth. This is already apparent in the livestock sector in the Netherlands, where phosphate regulations imposed by the government caused a reduction in the number of cows farmers can keep. A perfect example of how an environmental limit is slowing down the dairy economy. There is, obviously, no doubt that science, technology and innovation are essential for the future of humanity. However, it seems that the technical achievements of humanity, particularly those backed up by commercial interest, are shaping our present world. This is instead of first visualizing a world we want to live in, and then using science and technology to realize it. There are plenty of educated people who want to be a scientist. We should, for instance, decide if they will use chemicals within agriculture instead of the associated companies currently making that decision. So the question remains as to whether we should continue on our current path of pursuing ever-increasing scale, moving further and further away from nature and becoming increasingly dependent on business conglomerates – or is there another way? In this respect, I want to point out two directions we can choose between – one based on technocratic technology and one on nature-based technology.

Technocratic technology

In the previous section, I addressed various conceptual directions regarding the solution of the environmental problems caused by humanity. The prevailing view is that it must be scientifically and technologically viable to solve the pollution and depletion problems of today and in the future. Such a vision implies a world in which an increasingly technocratic society emerges with elite technologies, such as nuclear power plants and oil refineries. These structures can only be invented, realized and managed by rich and well-organized countries. These advancements may not offer the best possible solution to certain problems, but are rather technologies from which money can be earned.

For example, carrot juice happens to be very healthy for our intestines, but carrot juice will not put a lot of money in the bank, let alone allow for carrot juice advertising on TV. Instead, expensive dairy drinks with probiotics for healthy intestines are broadcasted. In comparison, carrot juice appears old-fashioned and outmoded. Yet it is clear that carrot juice is much cheaper and the production of carrot juice requires much less water than dairy drinks. As such, it would be better for society if we drank carrot juice instead of dairy drinks. Scientists who want to research the effect of carrot juice on the intestines, however, cannot attract money for the research, because funding for universities is often linked to support from the industry. The industry wants research into dairy drinks, which is patentable and where extra money can be made, so the scientists will duly study the effect of dairy drinks on the intestines. It is hardly surprising that we have created a world where the most technologically challenging solutions are supported only if money can be earned or a market can be won. Companies owe it to their shareholders to optimize their profits, and in doing so, they only have to comply with the law. Moral and ethical considerations must not stand in their way.

For example, old and resistant cereal varieties have been replaced by genetically manipulated grains that are bred in such a way that they no longer yield fertile seeds. This provides better control over the farmers, who are unable to produce their own seeds and will have to buy them anew every year. This is a way for companies to create new markets and gain more revenue. In the same vain, companies are facilitating interplanetary space travel, in the first phase for rich fare-paying passengers, but with the ultimate aim to flee into space if life on Earth is no longer possible. Blue Origin, owned by Jeff Bezos, the founder of Amazon.com, is such a company. Isn't it enough that we already have our hands full with the problems on Earth? The idea that we can flee and move to another planet is, of course, nonsense. There is not sufficient material and energy on Earth to build enough rockets. Neither does collecting raw materials from another planet hold much promise, given the very small weight a rocket can carry compared to the raw materials it will cost to transport these.

Nature-based technology

Numerous technocratic developments often involve risks that are not assessed sufficiently. Many factories, medicines, biological modifications, etc., turn out to be harmful at a later stage. For example, the danger of radioactive radiation was underestimated when thousands of soldiers in Nevada went on maneuvers in that area after nuclear explosions. Many pesticides have been banned for their unexpected effects on the bird population, for example. All too often products which are recognized by science as safe, subsequently turn out to be dangerous because it simply proves impossible to get a total grasp of the complexity of natural processes.

The complexity of the human body, ecology, the universe and indeed of all processes in nature is tremendous. Wouldn't it be more logical to develop technologies that are better suited to, and interchangeable with nature? Advancements that will make use of precisely this infinite complexity of natural processes, as opposed to developing the technologically highest achievable option? This vision, which I call 'nature-based technology', aims to use science and technology to solve societal problems, employing the most creative innovations available. These innovation processes will not only be reliant on profit-driven targets and controls of conglomerates. These companies are of course free to innovate, but their support will no longer be a prerequisite, as it is now. These natural innovations do not have to be techno-logical masterpieces, and the fact that they can also be used by poor countries is a big plus. An absolute requirement is that the technology must be sustainable, so it must not cause pollution or deplete resources further. However, this requires a new collaboration and approach from develop-ers. In this situation, we will strive for what is most optimal for society, rather than doing what is most beneficial for the shareholders. The innovation processes should, therefore, be co-directed by societal organizations and perhaps by the government.

Obviously, this demands a major change in the direction of our innovation processes. This includes our current innova-tion system, where the organization is rather peculiar, in-cluding substantial lobbying forces. For example, let's take a look at where the manure for Dutch agriculture comes from. There are 2 million hectares of farmland in the Nether-lands available for fertilization. The amount of manure we have at our disposal consists of human waste (or sewage sludge), cow, pig and chicken manure. We also use artificial fertilizer from factories. At this moment, with our livestock manure and the sewage sludge of the Dutch population, we can supply more than four times the amount of nutrients needed for our agriculture. Nevertheless, artificial fertilizers are not banned in the Netherlands and are still being used. In fact, farmers are forced by law to use artificial fertilizer because they are only allowed to use a certain amount of animal manure, and no human manure at all. Therefore, we have to burn our surplus of sewage sludge – human excre-ment – and chicken manure, and on top of that, we are also transporting huge amounts of animal manure across the border. By burning the manure, it loses its nutrients and the burned ashes are disposed of. Wouldn't it be more logical to develop technologies with which nutrients from human and animal manure can be generated? Our surplus could then be exported to countries that have too little. This would also put a swift end to the use of artificial fertilizers: the global nitrogen fertilizer industry already accounts for 100 billion euros per year alone, and uses no less than a gigantic 2% of all fossil fuels. More than two-thirds of the nitrogen spread on agricultural soil ends up as either air or water pollution. This can easily be prevented by reusing animal and human manure – a perfect example of nature-based technology: technology that characterizes itself by working with nature instead of going against it.

Visions for the future

Water and energy

A solution to climate change based on technocratic technology would be to simply keep burning coal and oil, and to store its CO2 emissions underground. This technology, however, costs a lot of energy and, moreover, our sub-soil can only globally harbor 2% of the CO2 that we emit. Nor does this solution take into account the continuing non-CO2 air pollution resulting from the incineration of fossil fuels. Last, but not least, this technology costs a lot of extra water on top of the amount already used by the oil and coal plants.

An alternative is the transition to nuclear energy, which sounds like a good idea to a lot of people. However, this technology requires steam, which demands large amounts of fresh water that will simply not be available in the future. What is more, if we look at the devastation this technology has brought about in Chernobyl and Fukushima, areas that have become unsuitable for agriculture and, for that matter, uninhabitable for tens of thousands of years, one may wonder if this is the right way. In the water sector, much is expected of nuclear fusion, because it is often said that by using nuclear fusion there will be enough cheap energy to desalinate seawater. The question remains, however, whether this will help, because the agricultural areas that need the water the most are far away from the sea. The fees for transporting massive amounts of water to these places will be enormous.

It is also easily overlooked that the cost of energy for desalinating seawater is only a small part of the total sum. Most of the finances will be absorbed by the high-tech installations themselves. Even if we should get energy for free, seawater desalination will always remain very expensive and will not become much cheaper than by 30%, at maximum. Another drawback is that desalination technologies will also pollute the coastal seas in the process, as the waste streams are heavy (from the very high salt concentration) and contain special chemicals, which harm life at the bottom of the sea. Desalination of seawater is used on a large scale to provide cities with fresh water, but for the development of basic agriculture, for wheat, rice and corn, this technology will not provide a solution.

Instead, we should look for technologies that are closer to nature and have no negative consequences for the environment. These nature-based technologies also require high-level scientific research and innovation, and can also be high-tech. An example would be planting trees on devastated and degraded land where agriculture is no longer possible. Reforestation requires high-tech solutions. It is evident that cutting down forests has enormous consequences in the form of droughts and floods. Trees can retain raindrops and gradually release them into the soil and underground reservoirs. Without forests, the rain immediately flows away without being absorbed into the ground. As a result, the water flows downstream far too quickly, leading to floods. In those cases, the chance of drought is also much more eminent, because the soil does not contain any water reserves to be released slowly in seasons with no rain. In addition, forests can attract clouds that are formed above the sea and, just as important, forests transport this water further inland by evaporation. In this way, the water is used several times across several forests. Every forest is transporting the water even further. By cutting forests at the seacoasts of continents, the water originating from oceans can no longer be transported further inland. In order for the forests to have this unique quality, they require at least 50 trees per hectare. And for this 'pull factor' to take effect, it is necessary to prevent a forest-free zone of more than 300 km between the continental area and the ocean. The trees can thus attract clouds from the sea, allowing the clouds to rain over land, catching the rain with their leaves and releasing the water into the soil and/or transporting the 'ocean' water inland by evaporation through their leaves. In this way forests can prevent both droughts and floods, thus producing extra fresh water that can be used for agriculture. Mankind has, however, already cut about 45% of the primordial trees – often along the coasts – causing many regions to suffer from droughts. By planting trees and putting these natural rain pumps back into operation, we can restore fresh water flows, and will no longer be dependent on groundwater or seawater desalination. Equally important, these trees will also capture CO_2. All the CO_2 that we are currently emitting can be stored in new forests planted on the unimaginable amount of 2 billion hectares of degenerated agricultural area that is available for reforestation. The cost for CO_2 reduction by nuclear energy is about 90 euros per ton, and for underground CO_2 storage it amounts to approximately 25 euros per ton. Planting forests on wasted agricultural land will capture CO_2 for less than 5 euros per ton. In addition, it will also provide more fresh water in dry areas and prosperity for the poorest regions, instead of radioactive waste in the case of nuclear energy, and no positive effects on the rest of the world in the case of underground CO2 storage.

In turn, these new forests will enable agricultural production as well, as this offers the opportunity to plant carefully selected and useful trees. As a result, the income of the local population will increase, which is all the more important as degenerated agricultural areas are often found in the poorest regions of the world. Finally, this new water flow will affect the sea level. At present, 25% of the sea level rise is caused by pumping up fossil groundwater. This fossil groundwater has been stored underground for millions of years and is now used for irrigation. This irrigation water subsequently ends up in the sea through evaporation, and in such large quantities that the sea level rises as a result. Thanks to the newly planted forests we can put an end to this fossil water consumption, since the trees will capture a tremendous amount of water in the soil and underground reservoirs, which will reduce the rise in sea level.

There are several ways to restore these forests. I would like to mention two, of which I will discuss the Groasis Waterboxx first. This high-end technology has become a Dutch National Icon in 2016. The Dutch government grants three National Icons every two years to top innovations that can be of great importance to the world. This technology consists of a cardboard box with a hole in the middle, through which a tree is planted. The box is filled with water and through ribbons (capillaries) this water can be fed to the roots until they have become long enough to find water in the soil themselves. In principle, one filling of the Waterboxx should suffice. Any rain will also end up in the box. For very poor soils, the box sometimes needs to be refilled. Therefore, there is no need to construct irrigation channels. Another method has been developed by the Tree-to-Be foundation. For this technology, degenerated grounds are selected where there are still some trees left. These trees must be accustomed to the natural conditions in the area. They are stimulated by a special pruning method to restore their root systems. In this way, these ailing forests are brought back to life. This is a very cheap method.

Maybe it is time that we choose in what kind of world we want to live. In a world that continues to engage in nuclear energy and to store the CO_2 of coal and gas underground, both at huge costs? A world where we pollute the coastal seas in order to desalinate the water? Or a world where we restore forests using degenerated land, solving both the CO_2 and water problems and creating prosperity in the poorest areas?

Phosphate

Phosphate is a crucial molecule for humans and animals. Without phosphate, life as we know it would not be possible. Phosphate is an important part of all living beings: of bacteria, plants and animals as well as people. We as humanity are currently unable to keep the phosphate that we use in the agricultural cycle. After all, phosphate mainly ends up in the excrement of humans and animals, and resides in the remains of plants. However, because we over-fertilize, destroy manure and let a lot of phosphate disappear into the coastal seas; a lot of phosphate is lost. Once in the sea, phosphate can no longer be procured. This means that most of the phosphate must be extracted from mines. This phosphate rock contains dangerous metals such as cadmium, which enter our food via artificial fertilizers. More importantly, we estimate that all phosphate mines will be depleted in about 300 years. Some refute this problem by saying this could take up to 500 years, or maybe even longer. Either way, I do think it is of vital importance, as the Earth will remain habitable for another billion years before the Sun gets too hot for life to survive. From that perspective, 500 years is only a fraction, and we should do everything we can to ensure that all the phosphate we use is fully recycled. Consequently, we will have to put a stop to over-fertilization and carefully manage all phosphate streams, such as human and animal excrement. As phosphate also pollutes the lakes and coastal seas, it should no longer be discharged into these waters. As mentioned before, we are currently burning all phosphate in the Netherlands from our human excrement. There are also plans to gasify pig manure. These are all processes in which the natural cycle is disrupted.

It would be much more natural and eco-friendly to fertilize with reused human and animal excrement instead of imported mined phosphate. These human excrements will again become part of the natural chain and be used by agriculture, as was customary until the Second World War in the Netherlands. I still know Dutch people who remember the 'barrel system', where the human excrement was collected and picked up door-to-door, to be used directly in agriculture. This yielded good results; however, this system has been replaced by the toilet and sewage for the sake of luxury and hygiene. Sewage treatment became necessary because a toilet mixes the droppings with water for flushing. In waste treatment, the excrement ends up in the sewage sludge after biological treatment, which is then used in agriculture for many years, and is often still used in agriculture in many countries. In the Netherlands this is no longer allowed and in other wealthy countries it is discouraged, with the most important underlying idea that at some point densely populated areas are faced with an oversupply of animal manure because of their intensive bio-industry for pigs and chickens.

The alternative use for our sewage sludge is incineration. This sludge combustion not only leads to the loss of all phosphate, but also to the loss of all micronutrients as organic sewage sludge contains essential micronutrients. Micronutrients are molecules essential for plants, animals and humans. We need about thirty molecules, of which several are scarce on Earth. For instance, compounds such as zinc, copper, iodine, boron, selenium, cobalt, and nickel. These compounds are available in the correct proportions in sewage sludge, and by re-applying the sludge in agriculture we can prevent shortages elsewhere in the world. By using fertilizer instead, nutrients do not end up in the proper proportions on land, which will inevitably have consequences for the health of the plants, but also for the food. Due to the manure surplus in the Netherlands, sewage sludge is not used in agriculture and so the Dutch society, and many other societies, appear to disrupt the global nutrient chain. Worldwide, the nutrient flows are out of balance. Densely populated areas import animal feed on a large scale, which contains a lot of phosphate. As a result, the countries where the animal feed is produced are facing phosphate shortages and have to procure phosphate from the mines. Conversely, there is far too much phosphate in the Netherlands and similar densely populated areas, through imported fertilizers and animal feed, yet we destroy the phosphate in the sewage sludge. The Netherlands is an example of a mega destroyer of the world's precious phosphate reserves. In addition, the sludge contains a lot of organic matter that is an important part of agricultural land. Many soils benefit from a higher content of organic matter, because this increases plant fertility and protects the soil against erosion. Additionally, a higher content of organic matter in the soil helps to capture more CO_2 in a natural way. Reducing the import of animal feed from areas with water and nutrient scarcity and putting a stop to destroying our own phosphate and micronutrients excesses will prevent a lot of misery in many regions of the world.

Applying sewage sludge in agriculture is legally prohibited in the Netherlands, in particular via the heavy metals legislation. For a sustainable solution, it would be interesting to investigate in what ways our sewage sludge could become cleaner for agriculture. Separate collection of toilet wastewater can produce much cleaner fertilizer than we are achieving now with sewage sludge. To do so, the toilets will have to be disconnected from the sewer to collect the excrement flow separately. Tests in Sneek, a Dutch town, have shown that human fertilizer from toilet water alone is much cleaner than average sewage sludge, and is easy to reuse in agriculture. This is because it is not blended with household chemicals like detergents, sunscreen products, shampoo, etc.

Thanks to our current import policy of animal feed there is already a surplus of nutrients in the Netherlands. A solution would be, instead of importing animal feed from other continents, to obtain the animal feed from a region to which we can deliver our organic fertilizer. For instance within a distance of hundreds of kilometers around the Netherlands, provided we can make a dryer manure product. In addition to the animal manure, we must also include human manure, which in the Netherlands has now fallen away as a supplier of nutrients. In this way, the Netherlands can stop destroying phosphate.

Agriculture

In agriculture, increasing production is still the main goal. This is achieved with many chemical pesticides, artificial fertilizers, big data and genetically modified plants. Because of the enormous amount of land devoted to agriculture worldwide, it has an immense impact on biodiversity and nature. Traditionally, no monocultures were used. There were many farmers who worked their small acreage for centuries and adapted to the possibilities of their land, mostly combining cattle and arable farming. After the Second World War, mechanization and chemical fertilizers paved the way for monocultures – making larger fields possible – and pesticides to control pests and diseases, which emerged when ecological boundaries of the original smaller plots were exceeded.

Artificial fertilizers also played an important role, making it no longer necessary to combine agriculture with holding livestock. Fertilizers and monocultures reduced the biological

Strip-cropping, Wageningen University (photographer: Dirk van Apeldoorn)

resistance of the crops and the soil. As a result, increasing amounts of pesticides are needed. In order to use pesticides more efficiently, genetically modified crops are also used, but mainly outside of Europe.

However, instead of increasing production, mono-agriculture appears to be reaching its limits, through resistance to pesticides, depletion, salination, destroyed soils and empty groundwater sources. Moreover, pest insects and weeds have become immune to the measures enabled by pesticides and genetic manipulation. Furthermore, beneficial insects that could keep the insect pests under control are disappearing at an alarming rate (75% reduction in Germany in the last decade). Researchers from NASA in the USA and other researchers from Japan have already designed robotic bees that can take over the role of bees to pollinate! Do we really want to eat genetically engineered crops from dead soils where insects and birds can no longer live and robotic bees are responsible for pollination?

New research shows that the old method of using mixed crops can be reintroduced by strip-cropping. These strips have the breadth of modern agricultural machinery. Mixed cultures appear to have 20% more yield than monocultures. For example, by growing wheat and peas in strips next to each other, the number of natural enemies of plague insects affecting the wheat, increases from 4 – from monocultures – to 16. The more biodiversity a crop brings with it, the better that crop can withstand diseases and pests. Woodlice, beetles, hoverflies, bees, and spiders form this biodiversity, which also continues in the soil with worms, nematodes, bacteria and fungi. A specially developed flower strip can be laid around the strips to attract the beneficial insects that will protect the wheat. During harvesting, insects can flit from the pea to the wheat lane and vice versa. Research into this method of agriculture is much more complicated than in monocultures, and much investigation is still needed to introduce this for all crops. But as a result, instead of boring monocultures with a lot of pesticides, we will soon get nice mixed cultures with increasing biodiversity in terms of insects and associated birds, with hardly any or much less chemicals.

Modern chemical agriculture is promoted as science-based and suggested as the only way to feed the world population. However, the majority of our agricultural production at this moment serves as animal feed. Let us look back at the unbelievable natural production of the prairies in the USA – before we started plowing the land. There were hundreds of different plants, with roots as deep as 10m with which they were able to withstand droughts. Approximately 120 million grazing animals, of which 60 million were bisons, could live on this fertile land, without any irrigation, pesticides, artificial fertilizer or machines. 'New' farming methods are trying to simulate this natural production, but now mainly for cattle (Harvey). These farmers no longer use pesticides or fertilizers, and, surprisingly, their cattle are so healthy they do not require medical chemicals anymore, such as vaccines and antibiotics. Production has soared and the organic content of the soil has increased fivefold within just a few years, as the soil is no longer opened by plowing. The profit for these farmers is much higher, because the operation is much more simplistic. Instead of plowing, irrigating, adding chemical supplements or pesticides, harvesting, and transporting to the stables, the cows will find their own food, without the need for modern chemical agriculture. As about twice the amount of carbon in the atmosphere can be found in the biomass (plants), and the organic matter in the soil contains even three times the air amount, using natural herding of cattle therefore can significantly reduce the CO_2 concentration. It will also reduce the water need. As such, not many will disagree that the 'natural' cattle products will be much more beneficial for our health than the 'chemical' ones.

Health

Modern healthcare is relying more and more on pharmaceutical chemicals, which have been detached from their infinitely complex natural context in order to apply them to the infinitely complex situation of our body. For example, patient care is greatly simplified to the effect of one chemical agent, such as the use of antibiotics, chemicals that are used to control bacterial infections. They are used indiscriminately. Dangerous bacteria can become resistant to these antibiotics. Some antibiotics cannot be used at all, because they no longer work. Moreover, there is the antibiotic resistance that now threatens our hospitals on a large scale, which is life-threatening for weak patients. It is expected that more than 10 million people per year will die due to antibiotic resistance worldwide. Previously, the excrements of sick people in hospitals were cooked before they were thrown away. Now everything goes into the sewer, such is our trust in antibiotics. As it turns out, this trust is misplaced. We may already be accustomed to very sick and old people dying from infections due to antibiotic-resistant bacteria, but suddenly this can also happen to young healthy people. An increasing number of people carry resistant bacteria. Once someone is infected with such bacteria, they can no longer be helped in a hospital. Just as we could die of a bladder infection in the past, this can also happen these days. It is of utmost importance that these dangerous resistant bacteria are prevented from spreading. This problem is even more acute in the slums of large cities, with very unhygienic living conditions and the large amount of antibiotics that are available to these communities. Here antibiotic resistance emerges on a very large scale, taking on epidemic proportions.

It is urgent we explore how to fight infections without these chemicals. An example is the use of an herb mixture in pig farming. Slime disease is a notoriously dangerous and sometimes fatal disease for pigs for which many antibiotics are used. It appears that this disease can also be prevented with an herbal mixture. It is cheaper and produces better results with regard to the disease, meaning fewer pigs will die. However, it will be quite a challenge to get this herbal mixture registered as a medicine. Registration is extremely expensive and the big question is: who will pay for it? It is very likely that it cannot be patented and is therefore not interesting for commercial organizations.

In the Netherlands, no research is done into herbal medicine any longer, and definitely not for human consumption; it is no longer part of the pharmaceutical faculty. This means there is less room for naturopathy in the Netherlands, even if it offers a more favorable course than mono-chemicals, especially for chronic diseases. Whereas agriculture has embarked on a new course through organic farming, naturopathy in healthcare finds itself under enormous pressure. It is not taken seriously and marginalized at every turn. They are also being slowly but surely squeezed out of insurance policies. Media are also often negative about naturopathy. Still, research into the relationship between psychological complaints, toxic load, healthy nutrition and chronic diseases, as advocated by naturopathy, is of utmost importance. In the cases of chronic illnesses, we are sentenced to taking – sometimes very heavy – medicines for life when treated by pharmaceuticals alone. These chemicals are not a cure, but only a way to constrain the symptoms. If these complaints could be alleviated by naturopathy, a greater focus should be placed on such an approach; good results have already been achieved with chronic diseases such as rheumatism, MS, fatigue syndrome and diabetes. So do we really want to be dependent on chemicals (medicines) for life, as is already the case for 7 out of 17 million Dutch people?

Another direction in technocratic technology is the development of advanced futuristic biotechnology, such as the use of stem cells to grow new organs, – for instance a new kidney to replace an old kidney – or to select or improve embryos. Furthermore, there are developments in nanotechnology and precision mechanics, among which a device that can be implanted in our body to replace our kidneys, or an exoskeleton to make soldiers more robust. Medicine is increasingly heading in the direction of the improvement of the (rich) human being instead of healing the human being;

this idea has already become common practice in plastic surgery. We will soon be seeing this in all sorts of areas. It goes without saying that these developments will only be available to the world's wealthiest, given the fact that we are not currently able to vaccinate all the children in the world. Do we really want to breed a new super race for the happy few, a race that looks beautiful thanks to expensive plastic surgery and can become very old thanks to expensive bio- and nanotechnology?

The 'makeability' illusion

In short, we can define technocratic technology as a way of production that is increasingly remote from nature. These forms of technology have a common denominator that they become increasingly large-scale and progressively more un-natural. They are based on the vision that human technology can ultimately control nature. Unfortunately, there are nu-merous signs that this is not going to happen. Yet insights that emphasize control over mysteries – such as the origins of our universe – are much more popular than concepts show-ing that solving the mysteries will never happen. This can be seen as an indication of our desire for a makeable world.

A certain complacency seems to have seeped into our minds, as we are shutting ourselves off for things we cannot know. Risks are no longer acceptable. In addition, this vision hangs on to the illusion that 10 billion people can all become equally rich, and that there are no limits. What is remark-able about these technical innovations is that they go to the edge of human ability, such as nuclear energy and genetic manipulation, but also that they are surrounded by major economic interests. There seems to be a tendency that pre-cisely this type of research is stimulated around the world, not because these technologies serve society best, but be-cause they bring in the money and give us a sense of make-ability. Technologies with earning potential are supported by the business community. Through this support they receive more attention at universities. This seems to be the case in all economic blocks, like China, EU and the USA, as they all announced to invest heavily in artificial intelligence.

It is surprising that the Health Council of the Netherlands comes up with the following advice in this high-tech era: watch your weight, make sure you exercise, eat a lot of vegetables and fruit, eat little meat, drink little alcohol and don't smoke. My non-expert grandfather gave me the same advice some forty years ago. However, if we look at what is trending in magazines, TV programs and commercials, it is all about sweeteners in Coca-Cola Zero, rather than apples. It's about slimming pills and powders instead of eating more vegetables. It is about plastic surgery instead of quitting smoking. For the ideas recommended by the Health Council, there are no commercials on TV and hardly any research is being carried out on this subject.

For years, agriculture has only focused on higher yields. In the past, grain was sprayed once a year against mildew, whereas nowadays this is done no less than five times a year. The yield is higher, but over the years, the costs of the pesticides have become larger than the extra yield of grain. However we cannot go back, because the varieties that used to be resistant to mildew are no longer available, since we only looked at more yield and we assumed to be able to tackle the mildew problem with pesticides. In my view, an unacceptable situation has arisen for society. Using five times as many pesticides is both undesirable for nature in and around the agricultural areas, and for the health of the farmer and the consumers of the grain. Naturally, the chemical manufacturers will have no objection to this situation. What will be the next technocratic solution in the fight against nature? Will they develop better chemicals? Will they develop genetically manipulated grain? Or can we just revert to the old grain varieties and settle for a bit less production, but a bit more profit for the farmer and a lot less unhealthy pesticides?

The mere focus on everything that is technologically feasible in the extreme puts us on the wrong track. Genetic manipulation is quite a feat of technology. Yet is it really that the world cannot manage without this technology? With genetically manipulated seed, another part of the yield from the farmer is transferred to the high-tech manufacturer; on top of fertilizers and pesticides. These technocratic solutions prove to be counterproductive. They do not make us healthier, at best they let us live a little longer. They certainly use much more water than the old technologies. And they are elite technologies that are based on a world vision in which we do not have to share. The facts, however, suggest that we will have to share if we do not want to immerse the rest of the world in poverty and hunger. For example, there is not enough water in the world to cater for a Western lifestyle for everyone. A global vision proclaiming that we do not have to share does not seem compatible with the current Western water footprint.

However, we can also adopt a completely different view, namely that people must learn to embrace the great mystery from which we come. Mankind will never be able to trace the infinite chain of cause and effect, and therefore it is far more sensible to stay close to nature and seek our innovations there, creating a world which we are not destroying and in which we share fairly.

5

We don't see things as they are,
we see them as we are.

Anaïs Nin (1961)

Invisible consciousness

What are the possible solutions to the environmental crisis? Many people have already tried to sort this out. Among their ideas, I see four directions for a solution that are often put forward.

- First, the concept of back to nature, back to the rather idealized times of the hunter-gatherers. In this scenario, however, we cannot accommodate 7 billion people on Earth, let alone 10 billion. Hunter-gatherers are not as sustainable as they would seem. With 10 billion hunter-gatherers, the Earth would be destroyed in no time. It is actually quite an achievement of modern society that we can live with so many people – thanks to our organizational capacities and level of knowledge and consciousness.

- Secondly, the idea that the environmental problem is caused by overpopulation. People who advocate for this theory do in fact feel there should be far fewer people on Earth, generally implying fewer 'other' people and, in particular, fewer poor people. This is a dangerous, narcissistic view, shared by more people than we think, given the fascination for the many films and books about the eradication of all humankind, with the exception of a select group. Take for example a recent book by Deon Meyer – *Fever* – in which a prominent elite group of people spreads a virus over humanity and only vaccinates themselves against this virus. This mindset is one of non-sharing and low ethics. Especially considering that the richest 1 billion use 80% of everything (water, raw materials, etc.). The removal of 6 billion poor people will do little to solve the environmental crisis, as long as the rich keep wanting more and more. In that case, there won't even be room for 1 billion rich people in the near future.

- Thirdly, the vision that science and technology will ultimately solve the environmental crisis, with its previous successes as the main argument. This vision is mentioned most frequently in the approach to the environmental crisis. Technology is becoming our hope for a solution. We solemnly believe in science as the savior of our problems, whereas at present there is no realistic basis for this conviction. We cling to this belief so that we can maintain our longer life expectancy and personal prosperity as the highest goals in our society. In other words: continue exactly as we are doing now. Despite our great wealth and technology, global biodiversity has never improved in the last century and we still have no idea how to solve the water crisis, therefore I believe that this direction will mainly result in the West saving their own hides and leaving it up to the rest of the world to deal with the negative consequences. Indeed, models show that the water shortage will hit Europe and America to a lesser extent than the rest of the world. It is highly questionable whether the planet will provide space for 10 billion people in this scheme. It is forecast that in large agricultural areas the water supplies will run out this very century, so there will never be enough food for 10 billion people with our present diets and way of food production. This mindset is also an example of an unwillingness to share and with little world perspective.

- If we cannot go back to a world with fewer people and if technology is not the solution, all that remains is for us to change ourselves. To achieve this, a stronger feeling of mutual connection with one another will have to take root in a large number of people. In the process, people in the West – and rich people in other parts of the world – will increasingly realize the consequences of their actions for the rest of the world, engendering more solidarity. Egocentricity will decrease and with it our greed and envy. We will be able to join forces more effectively, resulting in less waste and a better management of the ecosystems. By giving less priority to economic growth for the rich, the Earth will easily be able to accommodate 10 billion people. This culture will embrace the higher values and will create a society in which personal growth of consciousness is important. Everyone in this society will be encouraged to strive for a raised level of consciousness, with special attention paid to children. As Gandhi said, 'There is enough for everybody's need, but not for everybody's greed.' With less greed and better cooperation, the world will easily accommodate 10 billion people.

How can we change, and what exactly does that mean? Many feel that not all is well with the world and our Western civilization is constantly plagued with self-doubt. But what actually makes a civilization? On what basis can we compare civilizations? Perhaps the higher values of a civilization might be a good starting point.

For example, regarding the environmental crisis, a focus on the future is very important. The Dutch have invested so much in their protection against flooding in the future, that the statistical chance of a flood is only once every 100,000 years. In comparison, in New Orleans the likelihood is at least once every hundred years. The Japanese have protected themselves against very powerful earthquakes. Their cities have been built earthquake-resistant. Whereas in Italy, even a new school building collapses in an earthquake. A future direction relies on the extent to which people want to contribute to a common cause, such as the environment. People create a vulnerable society if they only focus on their short-term desires, for themselves and their families, instead of working together as a society and investing in our common future.

In order to solve our environmental problems, we need a society that is jointly focused on solving problems in the future. Jared Diamond describes in his book *Collapse* a number of previous civilizations that have been destroyed by environmental problems. He compares, for example, the downfall of the Vikings in Greenland, as opposed to the Vikings in Iceland, who were able to overcome their environmental problems. A perfect example of the necessity of future orientation and cooperation.

In this day and age, we can add another concept to these higher values, namely world perspective. Nowadays the material flows have developed to such a scale, that what we are doing at one end of the world also influences the other. The behavior of the Vikings on Greenland had no effect whatsoever on the people in Australia. However, the hole in the ozone layer above the South Pole and Australia is mainly caused by the Northern Hemisphere using propellants in spray cans. Our awareness of the effects of our behavior on the rest of the world has now become essential for solving the water crisis. Arnold Toynbee has analyzed 20 civilizations that have perished, such as the Roman Empire. In his opinion, in all these cases, the loss of higher values was the cause of the collapse of these civilizations.

Where do these higher values come from, you ask? These stem from our consciousness. This consciousness of a person or a culture is invisible and is often overlooked in many kinds of ideas and decisions. In the 50,000 years that modern man exists, the average level of consciousness has increased considerably. When one reaches more consciousness, one views the world differently. When we look back at the Dutch East India Company era (17th and 18th century), a time with horrible and cruel death sentences and slave trade, we do so through different eyes – a different consciousness – than the people did at that time, and we cannot pass judgment on this past society. It is not fair to judge human practices from the past with less consciousness against our current higher consciousness. For we have very different values nowadays and we can, in fact, only judge with our current level of consciousness. At every higher level, man becomes less egocentric.

This is one of the most optimistic and hopeful ideas I know. In the past 50,000 years, we have become less self-centered on average. Visible consequences are, for instance, fewer and fewer deaths from violence. This also reflects the fact that we can now live with 7 billion people on Earth due to our higher consciousness. The ability to share with people outside our family or tribe depends on someone's level of consciousness. The individual levels of consciousness determine the quality and coherence of the collective consciousness. With a qualitatively higher and more coherent collective awareness, the inevitable notion of sharing will evolve more or less spontaneously (Van Eijk). An important goal for mankind is therefore to continue to work towards achieving more consciousness. Unfortunately, in the Western world consciousness is increasingly denied, let alone sought for.

There is no good definition of consciousness. Just like life, consciousness is something that is beyond our grasp. We do not know where it comes from. We cannot know it. But we can try to describe consciousness. Consciousness is just like insight, something we cannot see. Everyone understands that mathematical insight is required for solving a mathematical problem, and a higher mathematical insight for a more difficult problem. Where this mathematical insight comes from is not clear. Some have it automatically. Others will only be able to make sums after an exhausting number of hours practicing. Mathematical insight is invisible. We can only experience it. This also applies to consciousness and higher consciousness. Consciousness can only be experienced. Consciousness is something that we are not born with. We must all build our consciousness ourselves during our lives and one person may gain more consciousness than the other. To make it even more complex, it is thought that we have achieved different levels of consciousness in different fields of our personality. It also depends on the culture in which we are born. In a theoretical case, if someone is educated by wolves, this person will have a lower level of consciousness than when educated in a society with schools. We could simplify that every society has an average level of consciousness and most people will automatically achieve this level with a normal upbringing. Every society also has outliers – in either direction. After the Big Bang, it took almost 14 billion years before man came into being, about a billion years after the first life form emerged. Our future is not in further evolution, which goes far too slowly, but in our cultural development and the advancement of our consciousness. The time scale for this is only hundreds of years. What I want to make clear in this chapter is that the level of consciousness a person or a society has, is crucial to how that person or society views and interprets complex matters. For example, an atheist scientist from our current rational society will interpret the Bible very differently than a scientist from the Middle Ages. This invisible consciousness must be taken into account when we are looking into complex matters such as environmental problems, and more specifically the water crisis.

Stages of consciousness

Writing and thinking about consciousness is much more difficult than doing so about the facts and laws of science. Consciousness is invisible or, at best, only visible to the 'conscious'. Yet everyone has a consciousness, and there appears to be multiple stages of consciousness in our development from child to adult. We cannot see from others what level of consciousness they have reached and this is also incredibly complicated. To describe this, there are several models. The model of Ken Wilber contains the most stages of consciousness and that is why I chose his model as a guide. The basic idea behind this model is that we, both as humans and as mankind, can grow in consciousness, thus reaching higher stages of consciousness and thereby become less self-centered.

In this respect it is important that we recognize that there are higher and lower values. Without this hierarchy between higher and lower, stages of consciousness mean nothing. On the one hand, there is the vision that everything in the universe, including life and humanity, is the result of chance and evolution; man is nothing more than an accidental spontaneous bundling of atoms. If everything is a coincidence, it is difficult to recognize a hierarchy. In this 'It' view (see Chapter 3), everything is equally important. There is no higher or lower. On the other hand, many people fight for animal rights. They consider the life and welfare of animals more important than that of plants. Vegetarians eat plants without remorse, but animals are off their menu. Animals seem to be placed above plants. And apparently cutting down a tree affects people more than digging a pit. So, in turn, plants are above soil and stones. This creates a hierarchy: soil, stones and minerals are at the bottom. Then come plants, where grass is lower than age-old trees. Above that are animals, with insects being of a lower order than mammals and human-like animals such as chimpanzees. Finally, there are humans. The majority sees human life as the highest good. Schumacher summarized this in four levels: lifeless (minerals), life (plants), consciousness (animals) and higher consciousness (humans). In this hierarchy, higher consciousness is at the top. It is a hierarchy of complexity, with minerals having the lowest complexity and man the highest. Man incorporates all lower levels (minerals, life, consciousness and higher consciousness). This hierarchy is therefore very important to determine the value of humans. Man is the highest on Earth and, as long as we do not discover higher consciousness on other planets, the highest in the universe. From this point of view, humanity is not a plague for the Earth but rather a blessing, because it increases the amount of complexity, which is apparently the driving force behind evolution. With a growing humanity, we are increasing the amount of consciousness on Earth. This is perhaps the goal of man: to bring awareness to the universe. Within this human consciousness there appears to be a hierarchy between stages of higher consciousness. By not only increasing the number of people on Earth but also the level of consciousness per person, the amount of consciousness increases even further. The difference in these levels of consciousness is of great importance. It makes a lot of difference when we weigh the opinion of a slave trader or someone from a tribe of cannibals against the opinion of someone from our rational time with equal rights for everyone. The more awareness we have, the more important our contribution to a better world will be. According to Ken Wilber, some stages of consciousness can indeed be distinguished from one another. However, this is not as simple as it seems, as various aspects of our personality can be in different stages. This model is therefore a great simplification and serves only as an example. The stages that are important to this explanation are ascending to a higher level of consciousness, starting from magical, to mythical, rational and finally trans-rational or existential. In this scheme, consciousness increases across the levels. According to Wilber, the techno-economic organization of a society is associated with the

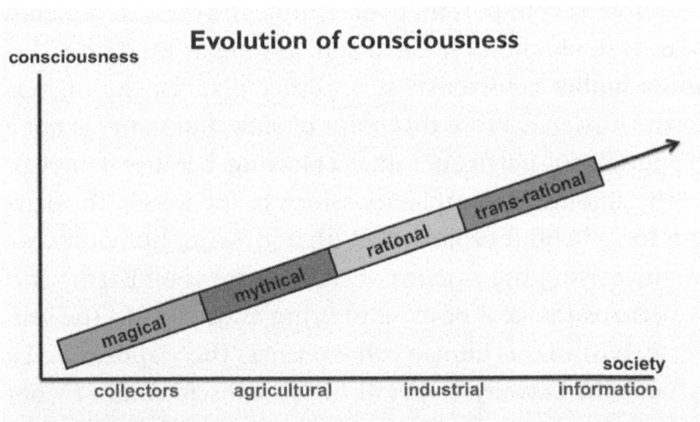

Consciousness has evolved over the past 50,000 years. Now the West is on the brink of a transition from rational to trans-rational (modified from Ken Wilber)

state of consciousness and the culture of the individuals within that society. To illustrate this, let us discuss the evolution of consciousness through the development of man.

Consciousness begins with every child at the lowest level: a baby is maximally self-centered and develops to the average of our (sub)culture. Every (sub)culture has its own average stage to which it has evolved. Ken Wilber has described a scheme in which the three segments (I, We and It) develop equally. In this context, I will focus on the techno-economic system (It), the stages of culture (We) and self-awareness (I). With self-awareness, Wilber's model assumes that consciousness will continue to increase during a lifetime. To be clear, consciousness is not the same as intelligence and high intelligence is not necessary for this development.

In the days of hunter-gatherers and shifting cultivation, the culture was *magical*. In these cultures people lived in tribes and families, and the average life expectancy was 22-25 years. Often this lifestyle is seen as ideal compared to our current stressful times. Hunter-gatherers had few possessions and little control over their environment. However, whether their society was indeed ideal, is a matter of debate. People were sacrificed and there was a lot of violence. With their consciousness, pictures and symbols could not yet be properly distinguished from reality, precluding the ability of writing. The idea that someone could harm a person by stabbing a needle in a doll representing that person, is an example of this magical thinking. Their way of thinking was strongly self-centered. Everything revolved around them personally. The clouds in the sky followed them personally.

This society was not forward-looking and communality was limited to the blood relatives or the tribe to which they belonged. If a certain area was unable to supply enough food, people would move to another area. There was no global perspective yet.

The gradual transition to the next, *mythical* stage of agrarian societies first took place in the Middle East. There was a change to agriculture, remaining in a permanent place and using a plow to till the land. By this, agriculture changed from a female task to a male job, as plowing was not possible for pregnant women. These agricultural methods ultimately yielded much more per person than the hunting and gathering did. This created the possibility that an increasing number of people did not have to deal with food acquisition, and the number of people also rose sharply. People started to settle in one location and would no longer roam as the land became occupied by farmers, who did not allow people to move around on their land. In this phase, the influence of man on the environment became much larger and the first environmental problems arose, such as deforestation and salinization. These processes degraded the land and caused major problems for societies. In this transition to large concentrations of populations with cattle, all kinds of new diseases rose. Excavations show that these first farmers were smaller and more susceptible to diseases than the hunter-gatherers. This transition has therefore caused many adjustment issues for humanity. Carel van Schaik and Kai Michel describe compellingly how during this transition the Bible was written in a period of a thousand years, with constant alterations. Many kinds of new life rules were laid down in this book, in an attempt to get a grip on this new situation. Great empires emerged during this time, due to the large armies that now could be formed.

The culture changed from magical to mythical. Consciousness no longer revolved only around ourselves. We discovered that we could not change the world individually, but instead we adopted an omnipotent God, whom we could please or ask things through commandments and prayers. Instead of performing a rain dance, which was purely about humanity speaking directly to the rain, we went to pray to God to ask for rain. God could make rain, which we no longer were able to do. In our thinking, it was important to conform to our social roles (in the Old Testament God can punish an entire population if only one subject does not keep to the law) and we especially relied on our families and our culture. Our thinking became conceptual. We could replace images, for example a drawing of a horse, with words. In Europe, it was the time of the Holy Mother Church and of the serfs. The seven deadly sins were described and the four virtues: sensibility (only he can do good who knows how things are and what their situation is), justice, courage and moderation (knowing when enough is enough). In the architecture of the cities, only public buildings and churches featured prominently; private buildings such as shops and banks played no role. Science was in the service of the Church in order to find the Truth and could not easily be misused – as opposed to our modern times nowadays, considering the enormous amounts invested in science for the arms race or for commercial purposes. The communality and future-thinking orientation increased significantly.

During the agrarian revolution, common irrigation systems were installed, ports were built and dikes were constructed.

In the era of the Industrial Revolution, the culture changed from mythical to *rational*. With this, the transformation of human power to machine started. As a result, physical strength for powering tools and machines became less important and gradually more equality between men and women developed – albeit together with the concurrent horrors of exploitation, child labor and pollution. This created the ideal of equality for everyone, with the abolition of slavery and a new attitude towards discrimination as a result. It was also the Age of Enlightenment.

Rather than drawing on the ideas from the sacred stories of the Bible, now our ideas were built on science. The 'It' perspective overpowered the 'I' and 'We' views, as in the ideas that everything had arisen by chance and through the spontaneous collision of random particles, that everything is only relative and does not serve a higher purpose. There is a scientificization of psychology and pedagogy. Personal interpretations are no longer acceptable and only what can be measured can be of value. The new goals in society are to increase the average life expectancy and the common standard of living.

There is a fascination with the new and modern, with the highest technical feat, such as fast transportation (the supersonic Concorde and currently the Hyperloop, a vacuum tube for super-fast transport between cities) and space travel. The 'It' idea that consciousness does not exist, that it is just a phenomenon of our brains and that, in fact, we are our brains, is gaining ground. Free will no longer exists. From now on, the idea of natural selection is that only the ones best adapted or the strongest will win. Listed companies are not driven by morality or ethics, but by the question of what is most profitable. Leaders of listed companies are expected to view profit as their highest goal, or else they can be sued in court. The only thing they have to adhere to is the legal regulations. The sheer size of the scaling-up of companies has the consequence that their actions have a wider effect on a worldwide scale. Coupled with our ever-increasing consumerism, in which our greed and envy are constantly aggravated by advertising, this will lead to a huge reduction in biodiversity, water resources and all sorts of environmental problems.

We are no longer prepared to muster up the finances to provide for our values, for example animal rights, while (Western) mankind has never been so rich. The ideal of the employee is to earn as much as possible in as little time as possible, and that of the employer to produce as much as possible with as few employees as possible. This increases productivity enormously, but more and more people are left without jobs. It won't be long until one country, such as China, can produce enough for the entire world. Estimates are that many people will no longer be needed for the economy. This will become a major problem, for work is considered by many to be very important for the development and sense of purpose of people. Some will say that this is the same as what was claimed in the 18th century in England, when the weaving looms were automated. They may have a point. Nevertheless, the question remains whether the new technologies are fundamentally different from before.

On the other hand, the increased knowledge and productivity, coupled with globalization, have enabled us to share our Earth with no less than 7 billion people. Personal freedom is another important rational idea. Civil rights and universal human rights are internationally recognized, especially in the Western world. Rational consciousness evolved into a world perspective. We are no longer just part of our own family or culture, but we see ourselves as part of a greater whole, as part of all people. People with a world perspective understand that decisions to buy something can also have consequences for people on the other side of the world. At this stage, people develop an autonomous ego. They form their opinions independently and no longer just adapt to those of their families, churches or politicians. Meaning, in the mythical world, was linked to religion. Serving God in order to become happy was the main goal. In these rational times, with an autonomous ego, this is a bit more complicated. We now have to choose a goal ourselves. For a strong ego, religion is just a choice. This autonomous ego is a prerequisite for further growth.

Finally, we arrive in the future, in the *existential or trans-rational* stage. Just as with the transition from hunter-gatherers to agriculture, the current growth in population and prosperity also demands a new growth of our consciousness. The rare level of consciousness that we will now have to strive for is the transformation from rational to trans-rational consciousness. In this stage of consciousness our ability to integrate dilemmas and paradoxes will increase. A paradox is connecting two notions that seem to be opposed. Carl Jung said that a paradox is an important means to come to the truth. From this, it follows that connecting opposed notions is important in ridding ourselves of our prejudices. We are facing many dilemmas, such as having to choose between a boring but sustainable and therefore more sober life, and an exciting life with all sorts of expensive excesses. In these dilemmas we can now let go of the more rational 'and/or', and connect these extremes by combining the exciting and sustainable. One example of a consciousness dilemma would be if we are concerned about climate change, yet continue to fly by airplane. In that case, some further self-examination is called for.

At this stage, we can observe our body, our thinking and our own feelings. We notice the observer within ourselves, the inner person who does not grow older, who always watches everything we do and think, and who understands that we are more than just our feelings and our thoughts. This observer, whom we often call 'I', can compare our thoughts from when we were five, to those at the age of fifty. With this observer-self, we can integrate body and mind, thereby strengthening our world perspective. We are now interested in what is fair and just, not only for ourselves and our own group or nation, but for everyone. My God is not the only God; my vision is not the only vision. My consciousness goes from the perspective of my group or nation to all of humanity. Here we can perceive the world soul in which all people are connected. At this stage, we see a world that needs care and mercy, and also our effort and dedication.

When we reach this stage, our greed and envy will diminish, we will rise above it all. A few percent of the population has already reached this stage (to Ken Wilber's estimation this applies to, for example, 4% of all Americans), a stage of development recognized by many psychologists. This concerns those people who act as global citizens and perceive that everyone is connected to one another. This consciousness also implies that we come to terms with our mortality and our mission, to find our own authentic place in the world. We can relinquish our symbols of immortality, such as futile attempts to become immortal by, for instance, leaving behind a work of art or a life's work. We can no longer ignore our responsibility by hiding in the herd.

Developing consciousness as a priority

In history books, the stages of consciousness are almost always overlooked when analyzing historical events and conflicts. The same applies to science fiction books in which the future is predicted, and also when we talk about spirituality. Despite the fact that consciousness does indeed make all the difference. With our rational view of equal rights for all and world perspective, history comes across as unrighteous with its serfs and slaves, but we cannot put ourselves in the place of the people who lived back then.

Since every child has to start from scratch, it is essential for the future that modern societies see the development of consciousness as the greatest priority. Not only do we have to acknowledge its various stages, but we also should stimulate its maximal growth. At this moment, we are divided. On the one hand, we have the conservative camp who believe in personal development and responsibility, but does not want to go beyond the mythical belief in an almighty God. On the other hand, there are the modern progressives who do not believe in an almighty God or even in anything transcendent as a higher consciousness, and emphasize a rational belief in material science and the optimization of economic systems and social conditions. If things are not well with the individual, they often attribute this to the social environment. Many believe we should seek a third way, in which our focus will be on the development of a trans-rational consciousness stage for our society. This vision of our reality progresses in the last three stages of consciousness from a mythical God to the rational mystery, and subsequently to higher consciousness. In such state of consciousness, we may feel a little anxious that we will lose touch with our reality, such as our contact with nature or with God, but that is not the case. On the contrary, we will perceive more clearly that nature is part of us and we will come to a better understanding of what we call – or used to call – God.

In the West we are questioning our achievements. We no longer know for sure whether we are indeed leading the way in the world. One perspective is that we are hanging on to the rational worldview that is set in our minds through ideas about competition and material explanations: man has been created by a random collision of particles and has no purpose on Earth. With such ideas, humanity can easily be viewed as a 'plague'. Conversely, we have the magical/mythical worldview that we have to return to nature or to our almighty Gods. Ken Wilber sketches a new, trans-rational perspective, in which we can continue to grow as a rational society. From this viewpoint, we see that our very high rational consciousness, with our autonomous ego, can continue to grow towards trans-rational consciousness, in which we can look further than our own country and ourselves. Our thinking will grow from abstract to visionary. We will become less self-centered and adopt higher values such as future orientation, communality and world perspective to shape our economy. We can then share with the people in the future, the people in our community and people in other parts of the world. Humanity will become a blessing. This perspective is much more inspiring and hopeful than the current one with its indifferent and narcissistic focus on the growth of prosperity. With the pursuit of higher consciousness, the world can get a new perspective. The growth of humanity to 10 billion people may very well depend on the development of our consciousness.

6

*If we feel intimately united with all that exists,
then sobriety and care will well up spontaneously.*

Pope Franciscus, *Laudato Si*

Autonomous strong ego

As already pointed out by many, the environmental problem cannot be solved with innovation and technology alone, a change in ourselves is also required. Therefore, this chapter focuses on the ethical roots of the solution to the environmental problem. Much has been written on this subject and – not unlike the orbit of the planets – we can only speculate on this through models, because the reality of human existence and consciousness is far too complex and largely unknown. The model of Ken Wilber describing the stages of consciousness (see Chapter 5) is therefore only a rough approximation of reality. It makes sense that the world looks different when you look at it from another level of consciousness. Everyone will agree that a child grows in consciousness as it matures. A young child, still completely self-centered, is convinced that we will not see him if he holds his hands in front of his eyes. An older child with more consciousness and a less self-centered view will realize that it does not work that way and that he really has to hide if he wants to be invisible. More surprising, though, is the idea of Wilber that even as adults we can grow further in consciousness. We can only see and interpret the world from our own level of consciousness.

Striving for as much personal prosperity and longevity as possible are the main goals of our current society. By personal prosperity, we mean that we can personally buy as many items and services as possible. Public prosperity is less popular: the less tax, the better. We strive to live as long as possible, in particular through good healthcare, which requires access to vaccines and other medicines, and also numerous hospitals. This is evidenced by the government policy of rich countries, which continues to strive for growth of the gross domestic product and then invests a very large part of this growth into healthcare. The richer an economy becomes, the more money goes to healthcare. The fact that Western societies are running into all kinds of problems based on this policy is becoming clearer. There are ever increasing signs of the fact that they are no longer on the right track. I need only mention the 7 million chronically ill in the Netherlands – out of a total of 17 million inhabitants – including the massive use of antidepressants, and the remarkable fact that more suicides occur in rich countries than in poor countries. The costs of economic growth are also constantly increasing. So the richer we become, the more losses we suffer. It is estimated that the rich countries and regions have already lost 18% of their prosperity to adverse effects of prosperity, such as social tensions caused by those excluded, investments in cyber security, depletion of agricultural land, shorter lives due to fine particulate matter from air pollution, and more.

One of the reasons for our current worldview is based on ideas that have been put into our heads by science. In the last fifty years, a specific concept for humanity has arisen. In this concept, people – like all other phenomena in the universe – are created through spontaneous cosmic statistics. As a result, we believe that humans don't really occupy a special position on Earth. Man is simply an intelligent animal without fur – also known as the 'naked monkey' – and a consequence of natural selection. Man is viewed as merely a phenomenon of physics, chemistry and biology. Or as some call it: only a few pounds of minerals and water.

People with this concept of humanity are somewhat suspicious about transcendental or mystical experiences, such as near-death experiences or feeling connected to nature, as they consider these experiences to be nothing more than little currents in our brains. In their eyes science can one day unravel the infinite chain of cause and effect, and society will then become malleable and our fate controllable. This way we can ultimately control nature and lead a full scientifically-based technological life.

This concept of humanity contributes to the narcissistic indifference that characterizes Western societies. As Pope Francis says: 'globalization is only the export of indifference.' And he has a point. Globalization is also a race to the bottom line, negatively affecting very important areas in our existence. This applies both to the labor rights of people and to environmental pollution. Labor-intensive work – such as assembling mobile phones or producing clothes in sewing workshops – has been moved to Asian countries, where people have to work for low wages in poor conditions. In the West, heavily polluting factories, such as tanneries, nearly disappeared; instead, we import the products minimizing the amount of pollution affecting us. Moreover, all new environmental regulations in the West are now weighed against competition with more lenient rulings elsewhere in the world. In this concept of humanity, what should man strive

View of the Milky Way (photographer: Jakob Woisetschläger, TU Graz/Wetsus)

for? If we are merely a product of a cosmic accident, there is basically no purpose in life and a sense of meaninglessness ensues. We can strive to live as happily as we can. For some this means living a hedonistic life: to party and consume as much as they can, in order to rise above the mundane everyday life. We hope to achieve this by becoming rich, or at least by trying to get by as well as we can.

From a rational perspective humanity can, however, also be viewed in another way. In my opinion, we cannot deny that there is a great mystery about the origin of the universe, life and our consciousness, and that it is not rational to expect that man can ever unravel this. Realizing the enormity of this mystery, we might ask ourselves whether evolution indeed has no direction and whether there is not a purpose for man. Love and intimacy are difficult to explain in a worldview of man as just a cosmic accident. How to explain the fact that – thanks to cruel experiences in American orphanages in the early 20th century – babies have a greater chance of dying if they do not experience intimacy with their caretakers? Based on these experiments, I conclude that intimacy is important at any age, even though we are not likely to die from lack of it immediately as we get older. Meaning, a sense of purpose, is also extremely important in life. Research shows that a sense of purpose is even more important for people than happiness, personal prosperity and longevity.

In another, not solely chemical, physical and biological concept of humanity, we can also recognize an inner dimension in which we can perceive our surroundings. Everyone can turn inward and observe this dimension. This inner consciousness, however, can also be experienced through established meditation procedures. However, this can only be done individually and therefore does not count as scientific. Verification is possible in consultation with others who also practice techniques for developing consciousness. Science has demonstrated the effects of meditation, proving that investing in our inner world has a real impact. Accepting the concept of consciousness and an inner world, an entirely new responsibility arises for man, not only material, but also an obligation to the growth of our own consciousness. We can create a shift of focus in our culture, from materialistic values to spiritual ones, and an increasing sense of purpose and meaning. We can transform into something higher, something that transcends us. In transcendent or mystical experiences, we have the feeling of rising above ourselves and we become aware of a higher reality. The insignificance of our short life against the infinite universe that exists – and will continue to exist for many billions of years – will for some underline the meaninglessness of our existence, and for others give transcendent inspiration. Inspiration that could lead to faith in an inner dimension and in higher consciousness. At the very least, there will be room for embracing the hidden logic behind things and phenomena. Depending on one's personal vision, this hidden logic can come from nature, God, fate, the inaccessible or other options. The aspirations from this concept of humanity will change our objectives from personal prosperity and longevity to a meaningful life and an honorable death.

Ken Wilber's model is indeed heartening in the sense that our current rational consciousness is already an elevated form of consciousness and that humanity has come yet again a step further. In rational consciousness, we no longer define ourselves by the group to which we belong, but by our own personality and ego. This appears to be a crucial point and is at odds with the idea that a strong ego impedes a growth of consciousness. According to Wilber, egoism is rather an indication of a lack of ego than too much, because a strong autonomous ego means more consciousness. An insecure ego will cling to property, prestige and power for security enduring the uncertainties of life. I am quite attracted to this positive and optimistic view of an autonomous ego, and that we, as humanity, are indeed still on the right track. This insight can shed new light on the next step for rich societies. The higher our consciousness, the less egocentric we become, and the more we can empathize with others. An autonomous ego is one condition for this growth.

Currently, there is little evidence of a higher consciousness that transcends our present rational one. This is because this form of higher consciousness is still quite scarce. In this respect, the mystics of the past can be our guide, according to Wilber. In his view, a mystical experience is not a nostalgic glance back in time, but a vision of how the future may look. Instead of a few people who used to be far ahead of their time, many more people can now reach this higher stage of consciousness. With all our present wealth and free time – and the many people with an autonomous ego – we have a much better starting position than the erstwhile classic masters. They have left us their legacy of prescriptions from

Christian mysticism, Hinduism and Zen Buddhism. Given the rareness of higher consciousness, it would be very useful if these experiences from the past were systematically tested and became publicly available.

This vision of our rational ego being a strong basis for higher consciousness can give our society a completely new purpose. This elevated awareness is our hope for a world in which we will become less self-centered and the crucial values of future orientation, communality and world perspective will increase. It is extremely questionable whether there can be a future without this higher form of consciousness, since the Earth cannot accommodate 10 billion people living the way we do now. In fact, not even 1 billion people will fit if we go on in this way. The first step, therefore, is to review the goals of our society. Let's trade in our current goals – trying to achieve happiness through luxury and personal prosperity, and immortality by clinging to life – and swap them for leading a meaningful life, and by choosing an honorable death.

A meaningful life

In recent years, social sciences have conducted research into how a person can lead a 'good' life. One of their crucial findings is that it is more important to live a meaningful life than a happy life. Meaningfulness and happiness may clash, but it turns out that meaningful activities ultimately lead to a deeper form of well-being. Emily Esfahani Smith describes multiple evidence that support this outcome, such as a comparison between the two. Happiness is convenient, feeling good with little stress and worries, having good physical health and enough money to buy everything we need or want. However, happiness is also linked to selfish behavior, 'taking instead of giving'. In contrast, a meaningful life is associated with 'giving', seeking connection and contributing to something that transcends us. Living a meaningful life means giving, caring and fighting for an ideal or conviction. For a meaningful life, we invest in matters that are bigger than just ourselves, ultimately giving more stress and worries than a mere happy life would. Research shows that there is no relationship between suicide and a happy or an unhappy life, while there is indeed a connection between suicide and a lack of meaning. 'Happy' countries such as Japan, Denmark and Finland have the highest numbers of suicides. The number of suicides is also increasing in the Netherlands. In 2016, no less than 1,900 people ended their lives, 550 more people than a decade ago. Poor countries, such as those in sub-Saharan Africa, score high on an unhappy life, as there is great poverty, but also score high in terms of meaningfulness – contrary to the rich countries – whilst having a much lower suicide rate.

In the USA, no less than a quarter of the population feels that their lives have no purpose. This is partly due to the current system of education, both in the USA and the rest of the developed world. The oldest universities, some of which are already more than thousands of years old, were established as theological universities, with the main aim of teaching their students how to deal with life. Emily Esfahani Smith describes how American colleges and universities were founded for this purpose. Students were introduced to theology and the classics. From the 18th century, the humanities took over the role of religion. Masterpieces of literature and philosophy, such as by Homer, Dante and Shakespeare, replaced the Bible in order to find the meaning of life in an organized way. In the 20th century, the situation changed with the arrival of research universities. To conduct research at these universities, separate studies were developed with their own objective methods. Students became highly specialized. The subjects that focused on the humanities disappeared gradually from the curricula. The focus on research replaced the idea that a meaningful life could be learned at a university. Many professors now do not consider 'meaning' to be an aspect related to their specialization and therefore do not find it appropriate to guide students in this subject, and may even find it embarrassing. The duty of a university nowadays is to teach students to think critically and to learn a certain subject, and not to guide students in the development of their personality.

Unfortunately, we may never fully understand the essence of life and therefore everyone will have to search for their own meaning, their own sense of purpose. The idea is that meaning will emerge when we connect our intentions and activities with something bigger than ourselves. That is why work is very important to many as a source of meaning, and, conversely, unemployment a disaster. Robotization and automation are likely to destroy a sense of purpose for many. The prediction that the majority of jobs as we now know them will be taken over by artificial intelligence in the next 20 years, gives many people the shivers. Emily Esfahani Smith mentions other examples of meaning as observed in her research, such as having ties with friends, family and partners. This is becoming increasingly difficult in this digitized world, where everyone spends an increasing amount of time on their smartphone instead of in direct contact with friends and neighbors. Other examples she encountered are: having the feeling that we matter in society or in the company where we work; having a purpose, such as doing something good for the world; taking care of our families; or being part of something bigger such as spirituality, world soul or mystery. These examples are applicable to everyone, regardless of their views on religion, science, or their concept of humanity.

An honorable death

We are increasingly inclined to hang on to life and in doing so we will leave a legacy of debt to future generations. In the Netherlands alone, the budget for healthcare is already reaching 100 billion euros a year of a gross national income of 700 billion. The government must implement all kinds of measures to keep this under control. However, why are we clinging so desperately to life? In one way it does seem logical, given the general view that after our lives there will be nothing and this life is all there is. Dying means the end of everything. From this point of view, we regard our death no longer a part of the mystery, but rather as the result of a technical defect. This means that we will use all sorts of aggressive techniques to extend our lives, at a high cost. These costs not only relate to the enormous fees of current practices, but also to the incredible investments we are making nowadays into medical techniques. We want to live as long as possible and see ourselves as the end product of a lifelong investment. The costs of extending our lives in the last years of life are just the beginning: many people are captivated by the possibilities to rejuvenate our bodies, and thus making a start on immortality. Medical science already predicts that it will not be long before this becomes a reality. For example, Google has set up a subsidiary company that focuses on conquering death. Google Ventures has already made several investments in companies with this mission. This endeavor also receives support from several billionaires. With the vision of man being a biological and chemical phenomenon, this approach seems logical. In addition, this puts our ego, which should last as long as possible, center stage, instead of our connection with family or society. This fragile and fearful ego, which has not yet made peace with our mortality, is now more important than the power of our society, which earlier generations died for. We will pass on our debts to new generations through our inclination to cling to life and our pursuit of immortality.

Our society, however, is not the first to regard death as the worst thing that can happen to a person. In the classical period, people did not so much fear death itself, but they considered a dishonorable, or even an ordinary death to be the worst thing that could happen to a human being. To classical heroes, our current most common end-of-life practice – namely, a serious disease (cancer, stroke, heart attack, etc.) followed by a period of struggling and being highly dependent on others for an average of four years – would be inconceivable. They would not want to live in a nursing home, with tubes in their noses, fully dependent on underpaid and stressed workers who have to see to their care.

We rarely see dying honorably in the service of future generations nowadays. Yet there are many people talking to their doctors about alternatives to aggressive treatments such as chemotherapy and heart surgery, in the final stages of their lives. Viewing death once again as being part of the mystery will support this approach. Today's society draws a cold picture: we consist of a bundle of dead atoms and death means that we disappear into the infinite number of atoms of the

Earth. However, nobody can say anything with any certainty about life after death. As mentioned earlier, life is something that transcends us. We do not know exactly what it is and where it comes from. This also automatically applies to death.

Would it not be more satisfying to see death as more than just nothing? A Dutch cardiac surgeon describes the near-death experiences of his patients. People who have had such an experience change their attitude to life and become less afraid of death. In *The Tibetan Book of Living and Dying* by Sogyal Rinpoche, the process of death and rebirth is described in great detail. There are different psychological and philosophical movements that study rebirth, and our Christian churches speak of the hereafter. These ideas, that death is not the end, are these not much more encouraging and comforting than the cold rational view that death is indeed the end? Since either view could be right, we may as well go for one of the more comforting concepts.

Suns and moons

For a development towards a fairer and more sustainable society, it is essential that good people take the lead and set an example. It has been scientifically proven, that people who have been treated fairly will be more inclined to treat other people, even strangers, fairly. So we can spread fairness ourselves. Together we create environmental problems by exceeding the carrying capacity of ecosystems. We can pollute the river Rhine to a certain extent, because the natural processes in the river will convert and remove this pollution. But if we pollute too much and exceed its natural capacity, all fish will die, as in 1970 when the Rhine was completely without oxygen. This principle also applies to water shortage. Nowadays nearly all people in the rich regions of the world use too much water and therefore we are all responsible for the depletion of water sources on other continents. This principle is called the 'tragedy of the commons', the commons being the common meadow of villages in the Middle Ages. As long as the capacity of this meadow was not exceeded, the sheep of the village could graze happily and the system remained stable for years. Until someone thought: 'Let's put a few extra sheep on our pasture.' And perhaps others thought: 'If he can do it, so can I.' The meadow became exhausted by overgrazing, and eventually disappeared altogether. The offenders enjoyed an additional advantage for some time at the cost of the others, because after that, the entire community was left empty-handed. Nearly all of us contribute to the global environmental problems, such as water shortages (we use more than 2,300 liters a day), the climate issue (we use fossil fuels), fine particulate emissions in the cities (with our wood stoves and cars), the plastic soup in the oceans (with our fleece sweaters releasing shreds of polyester), and many more.

Where does that put us in this tragedy? Do we, as an individual, want to be the cause, or rather the solution of these problems? Research shows that most people are neither good nor bad, but that they are easily influenced. They are, in fact, quite similar to moons. A moon cannot emit light on its own, this is only possible when illuminated by the Sun. In the same vein, the majority of people move with the prevailing opinions in their neighborhood. There are also people who are inherently good and just like a sun, they emit light themselves. The suns in our society, about 30% of all people (Peter Richerson, University of California) can therefore have a huge influence. These are the leaders we need, they cause other people to light up. Sadly, they are often unaware that they are suns. Nor are they always aware of what is good in our complex society. In our pursuit of a less egocentric society – which in my view is the only way to overcome the most serious environmental problems – raising awareness among the suns about their nature and position is therefore crucial. It is amazing how much impact a sun can have. Takuma Umoja moved from Texas to Jackson, Mississippi, an impoverished town, where only those not able to leave remained. Invisible people, in a hopeless situation, in the poorest area in the USA. He set up a cooperative and started to improve things, little by little. The residents thought Takuma would make no difference. Yet he persisted and kept on going. It took some getting used to in Mississippi, but it was contagious. After the first signs of improvement, the hopeless started to imitate him taking initiatives and becoming economically active again. Suns are positive and optimistic. As Martin Luther King said: 'Even if I knew that tomorrow the world would break into pieces, I would still plant an apple tree.'

In addition, the suns have another task that nowadays we all too easily tend to forget, namely keeping evil under control. Between 1% and 10% of people are malicious psychopaths. They can be compared to black holes. These celestial bodies can actively capture the light from the universe. Susceptible people (moons) will become very dark in the environment of these people. In criminal organizations, it is easy to see how much influence one leader can have on a lot of followers. During wars, the restraint on these psychopaths often diminishes and we can see what terrible things can happen. The Nazi regime taught us what can happen if psychopaths take over the official leadership of a country, incriminating an entire nation.

Neuroscience describes the Hubris Syndrome, in which people under the influence of power develop narcissistic personality traits in which they see the world as a place for their power play and personal glory. Symptoms of this syndrome are a sense of overconfidence in their own judgement and emphasis on their image. It seems that in our time, with high expectations of and room for business, we have given CEOs a lot of leeway, for example, leaders of large companies, who can implement immoral measures and take huge risks with society for their own gain. Take, for instance, the banking crisis as a result of incomprehensible financial constructions, or the oil spill in the Gulf of Mexico due to savings on security measures. It is up to the good leaders to keep these people away from such positions, or keep them under control. We observe every day what can happen when we do not adhere to this. If the management of companies is in the hands of good leaders, the market economy can contribute a lot to the world, for such leaders can seek solutions through free and creative market initiatives to the social problems of the world – instead of causing them.

Possession, prestige and power

In our Western society, greed is stimulated as much as possible. This is considered indispensable for economic growth. It is, however, becoming patently clear that this greed is currently one of the major forces behind the environmental crisis. How can we get this under control? Problems of our personality with regard to possession, prestige and power are timeless. In one way or another, we are insecure beings and try to defuse our insecurity through possession, prestige and power. In the Middle Ages, joining the monastery was seen as the best remedy against this trait. This drastic step involved surrendering all possessions by transferring it to the monastery, giving up all prestige by renouncing all titles and beautiful clothes in favor of a simple brown habit, and throwing aside all power by promising obedience. The underlying idea was that we can only control this addiction to possession, prestige and power by stepping out of society.

After the Reformation, the general view on this problem changed. We saw an opportunity to eliminate these addictions from our daily lives other than stepping out of society, by way of integrating our struggle against them into everyday life. This, however, requires self-control. It can be said that for many people today self-control is a rare commodity and, with respect to the environmental crisis, our addictions do get out of hand. Many people base their personality on possession, prestige or power or a combination of these.

This can take on such proportions that some people are psychologically lost when they lose their money or their job. They have become solely their property and lost track of the remaining part of themselves. These addictions affect every one of us. It is not as if power is only a problem for managers of companies and leaders of countries. Whenever we want to prove we are right, or are stubborn about something, it is all about power. Likewise, ownership is not just a problem for rich people. Whenever we appropriate something that is not ours, for example an idea or a public parking space, this comes into play. The idea that our partners, children and friends are ours is also a symptom of an addiction to ownership. Finally, the craving of prestige can also incite bad behavior. For example, in the academic world we see researchers who falsify or invent data for research papers, in order to get into the newspapers and acquire scientific prestige.

In order to do something about our addictions, we must tackle the underlying insecurity. I believe that there is a natural life path for everyone and that this path is different for us all. When people used to propose a new candidate to Napoleon for the post of general, Napoleon always asked if that person was lucky. Actually, he had the right idea, for what use is a general who is always unlucky? Such a general will not win many battles. Doing the things in which we have luck on our side seems a good compass for guiding us towards our unique life path. We can opt for fake security through possession, prestige and power, or we can choose our own path. This does not necessarily mean that possession, prestige and power need always be negative. For some positions, they are crucial, as long as we do not become reliant on them.

Hans Korteweg, a Dutch philosopher and psychotherapist, has labeled the force we have to combat in order to detach us from our addictions as the 'Guardian on the Threshold'. This Guardian is a perfect seller of our addictions. Our environment, for example, in terms of loved ones, family, friends or neighbors, can be the Guardian by keeping us in our current situation, by warning us that everything should stay the same for fear that when we take too many risks by taking new steps we may become poor or destroy our relationships. The Guardian can also adopt the form of fear for those new steps in our own thoughts. Such as not quitting a job that makes us unhappy, because we will never find a new job. We all have to cope with our own personal Guardian and must learn to deal with it in our own way. It requires self-insight and courage to choose our own lives and to let go of the addictions to possession, prestige and power.

Role of the Church

An important function of the Church, through all walks of life, is to provide a comprehensive system for classes in society to remind us of our commitment to become better people. What could be a better reminder of the seven works of mercy: feeding the hungry, giving drink to the thirsty, clothing the naked, visiting the imprisoned, helping strangers, sheltering the homeless, visiting the sick and burying the dead? Or of the four cardinal virtues: prudence, justice, courage and temperance? In church, we can also reflect on moments of joy (birth, marriage) and grief (disasters, funerals). We still see mercy in the food banks and other similar initiatives by churchgoers.

However, the main Churches in the West, especially in Europe, obviously do not adjust well to our ever-changing society. Secularization is advancing rapidly all over Europe. This decline in church attendance is contributed to the growth of the ego and the demystification of the world. That may be great news for many who see religion as a kind of superstition. In my view, it is bad news since nothing has taken its place to remind us to become better people. There is no longer a large-scale system where we can step out of our digitally isolated, egocentric, materialistic world on a regular basis. This very important role of the Churches has been lost for the majority of people. This inevitably leads to an increase in our egocentrism. We do not recognize anything higher than ourselves and see our ego as our new God. For many, the theory of evolution is also a reason to stop going to church, as it is not compatible with the Bible stories. Even with all our knowledge about evolution, the mystery remains undiminished, given the fact that we still do not know and probably never will know where life comes from. The Churches, however, could adopt a less conservative attitude towards the Bible, which has not been amended for more than 1,600 years, whereas this was common practice in the preceding thousand years. In an adapted Bible, evolution could very well have found a place, without losing the surpassing mystery.

The idea of an almighty God, who can allow so much misery in the world, is also an issue for many. However, it is impossible to form an image of God, who is in fact unthinkable and indescribable. A more mystical image of God that connects us all and that can also be an image of the world soul, will fit in better with our times. A high consciousness is needed to experience connection with the world soul ourselves. Brené Brown has defined spirituality as recognizing and celebrating that we are inextricably connected by a force that transcends us, and that this connecting force is based on love and compassion. Attending a church service together, singing together and wishing each other peace, gives us this sense of connection. At such moments we forget that we all believe something different and vote for separate political parties. We can also achieve this feeling with touching music performances and other such gatherings. The fact that people have become more self-centered, should in this regard better be rephrased to: people have acquired a more autonomous ego. We no longer automatically do what our families did within their pillar. We can now choose ourselves. The Churches have made an important contribution to raising consciousness. The question now is whether the Churches can regain their role in Western society by responding to the higher rational consciousness of this age. Rational consciousness is admittedly better than mythical consciousness, but in turn, trans-rational consciousness is superior to rational consciousness. It remains very important that people can strengthen themselves in the process of becoming a better person, among other things in order to keep our addiction to possession, prestige and power under control.

It would be a tragedy if the Churches were caught in conservative powers and – in a not too distant future – the last one would have to close their doors. The most beautiful churches can be found everywhere and in the most beautiful spots, and slowly but surely they are all converted for other uses. For example, before long there will only be approximately twenty open, active churches left in the main Dutch dioceses, compared to the hundreds we had 30 years ago. Wouldn't it be great if in Western countries with a rational consciousness experiments would start with a new Church approach that support growth to more consciousness?

7

Not everyone needs to become a world leader to change the world. Everyone can start in their own family or environment.

Kofi Annan

What can we do ourselves?

In this final chapter, I offer a number of practical tips that everyone can follow – without waiting for others – to start our contribution to a fairer and more sustainable world. Modest steps can have a huge impact, especially if they are followed en masse. The tips focus on increasing sustainability and consciousness.

Be aware you are a sun!
There are enough good people to reshape the world into a better place, however, many of them are unaware of their role, because they conveniently assume that everyone is the same as they are. Well, that is not true. The majority of people are easily influenced and are looking for a role model. The good people can therefore be compared to a sun. They can illuminate others. Suns have two tasks: first, to be an example for the many that can be influenced, and secondly to keep psychopaths under control. People who are easily influenced will not be able to keep psychopaths on the right track; they need good leaders, they need the suns. This is a responsibility that can have a lot of impact. It is unfortunate that many people are not aware of being a sun. Moreover, it is imperative that good people work together. Their effectiveness can be considerably increased by cooperation. One should develop a radar to recognize other good people. Deep inside, we already perceive this.

Eat meat like a lion

For about four consecutive days (out of ten), eat meat like a lion and no meat for the remaining six days (advice Nutrition Center: no more than 500 grams of red meat per week). And eat vegetables like a deer, which means every day (advice Nutrition Center: 200 grams per day). By eating less meat, we will reduce our footprint considerably and come closer to sustainable water consumption. It is also much healthier for our body. As far as eating vegetables is concerned, the effect it has for our health is reasonably well known. Eating meat only for a few days gives our body the chance to digest all proteins of the meat – including the bad ones – before we eat meat again. Even lions, who eat nothing but meat, need an average of 6 days to digest meat.

Walk a few kilometers every day

We are built to walk. It improves our immune system. By taking a brisk walk, our lymphatic system is stimulated. This is necessary because the lymph system does not have its own muscles and needs to be activated by physical activity. All in order for it to carry out the important task of disposing of our waste products. Sitting is the new smoking.

Tread carefully in the digital world

The digital world does not forget, and increasingly stands in the way of human contact. Once we get used to the Tinder dating system, we no longer dare to speak to anyone directly. Tinder will remember all our appointments and preferences now and forever. The latest trend is the Internet of Things. This allows devices to be linked wirelessly to the internet. Apart from the fact that the luxury of smart refrigerators and thermostats is probably not conducive for keeping our brains active, by connecting everything to the Internet or Things we also lose our privacy, and on top of that become vulnerable to unwanted influences when our 'smart' equipment is hacked. If entire container terminals can be hacked (see Chapter 2), this can certainly happen to our house. The devices connected to the Internet of Things are an easy target. Unsecured devices are, on average, attacked every two minutes. Last year, the Mirai worm infected over 380,000 devices. These devices were used to paralyze parts of the internet.

Read books

We get a lot more from books than from the newspapers and internet, which tend to sensationalize content. Even if a newspaper goes a little deeper, it is not always enough. A fragment from history, for example is out of context, whereas in a book the full context becomes clear. Experts frequently comment that a newspaper article about their subject was not completely accurate. Whereas, most of the time, books are written by the experts themselves. Additionally, reading from paper increases concentration by as much as 30% compared to reading from a screen. It may cost paper, but saves a lot of energy from displays and data storage systems.

Buy more expensive and more sustainable food

A vicious circle has emerged. Farm are becoming larger and larger in order to be able to become cheaper. This large scale has all kinds of consequences for the environment, due to the use of many pesticides (such as the fipronil in eggs problem in the Netherlands in 2017), fertilizers and antibiotics, increasingly larger machines and much less biodiversity. We see the loss of insects and birds, and the depletion of our water resources. Agriculture has a major impact on the global environment. However, by spending more on food, agriculture can become more sustainable. Organic farming is currently the only system that manages to charge more for sustainable products. Regular agriculture should also be able to charge higher prices for more sustainable products. This can start with us paying more for our food. That should be possible, as we spend a lower share of our income on our food than we did thirty years ago. This should be no problem since we apparently can afford to throw away 30% of our food, on average.

Support politicians who no longer take Western growth as a starting point

Further growth of wealth and consumption in the rich countries leads to an unfair seizure of the scarce water reserves, leaving others with too little. The West must grow in sustainability and not in productivity, and thus be a new example for the world. Growth in sustainability leads to a healthier living environment, and also many interesting jobs can continue to exist. Continuing on the current path of economic growth leads to further automation and inequality. Meaningful jobs, such as those of nurses and teachers, will become too expensive for average citizens and countless other jobs will disappear, as we are currently seeing at banks and insurers where computers are taking over. The environment will continue to deteriorate and biodiversity will steadily decline. We need to change our dream of ever-increasing luxury, because for many the opposite will happen.

Make time to be bored and to focus inward

Our inner world is at least as extensive as our outer world. Due to the enormous amount of stimuli from the outer world, however, we no longer have time to focus inward. With the arrival of smartphones and social media, this has become even worse. By occasionally consciously closing ourselves off from these stimuli, we have time for inspiration, time to become more creative, and to have more personal contact with our friends and family. Just put your smartphone away from time to time. Whenever we hear the familiar 'ping' signal, we check our phone, on average, within six seconds. This signal disrupts whatever we were doing, sometimes for no less than twenty minutes. Growth in consciousness also comes from our inner world. If we occasionally craft some silence by shielding ourselves against stimuli, we may find time for mystical experiences amidst nature. Invest in the development of your consciousness as an alternative to investments in luxury. For our children too, 'time to be bored' has become very important in this digital world, and being 'bored' while surrounded by nature can help them considerably.

Choose a purpose in life

In the end, we will get more satisfaction from a meaningful life. A purpose in life is very important in this respect. This can be anything: taking care of your family, caring for others, contributing to a better society, etc. If we cannot find an explicit purpose, why not take growth in consciousness as your purpose? That is crucial for humanity. This purpose can be pursued by everyone, both for themselves and by helping others in this area.

Determine what is enough as early as possible in your life

When we are young and starting our career, it is a good time to decide for ourselves when we have acquired enough. But also when you are older and have started your career a long time ago, it is important to determine what is enough as early as possible. If we can determine this for ourselves, it sets our minds at rest and offers an escape from the earning-spending prison. Every time people start earning more, the money runs out accordingly and a desire for more ensues. First, we make debts to buy a house. As soon as the house is paid off, we want a holiday home. And if we earn even more we buy a boat, or we want to go on a world trip. Or we want to outshine our neighbors or friends. If we have determined what is enough for ourselves, this is no longer an issue. We can say: a house, a car and one nice holiday per year is enough. From whatever we earn above our 'enough', we can invest in planting trees in devastated agricultural areas. This is how we can bring prosperity to the poorest areas.

Be generous, merciful and forgiving

Never see another person as an object. Neuroscientists have discovered that areas in our brains that normally light up when we identify objects, also light up when we no longer see our fellow human beings as human beings. This can be the case when we no longer see, for instance, opponents from another army – or even people from another football club – as people. Alternatively, people of other races, religions or sexual orientation can also suffer the same judgement. People who are the target of this dehumanization are depicted as being worthless, criminal or simply bad. In the end, these people are treated like things. I consider this to be a very low form of consciousness. According to Augustine, one of the early Church fathers, 'man can master his motives' – current scientists would call this his 'brain'. He stated that in order to lead a good life, it is important to learn how to fast and how to give. Ken Wilber says: 'let us grow in consciousness, so that we get a strong ego, with which we can withstand the addiction to possession, prestige and power.' Generosity, mercy and forgiveness are all forms of higher consciousness and, in our age, are more important than ever.

References

Barber, D. (2014). *The Third Plate*. New York: Penguin Group.

Bavel, B. van (2016). *The Invisible Hand? How Market Economies have Emerged and Declined Since AD 500*. Oxford: Oxford University Press.

Bergen, A. van (2014). *Gouden jaren*. Amsterdam: Atlas Contact. (Dutch)

Brown, B. (2017). *Braving the Wilderness*. London: Vermilion.

Diamond, D. (2005). *Collapse*. New York: Viking Press.

Eijk, T. van (2017). *Spinoza in het licht van bewustzijnsontwikkeling*. Lulu. (Dutch)

Harvey, G. (2016). *Grass-Fed Nation: Getting Back the Food We Deserve*. London: Icon Books.

Hoekstra, A.Y. (2013). *The Water Footprint of Modern Consumer Society*. Londen: Routeledge.

Korteweg, H., Korteweg-Frankhuisen, H. & Voigt, J. (1990) *De grote sprong*. Utrecht: Servire Uitgevers. (Dutch)

Paus Franciscus (2015). *Laudato Si'. On Care for Our Common Home*. London: Catholic Truth Society.

Piketty, T. (2014). *Capital in the Twenty-First Century*. Cambridge (MA): Harvard University Press.

Richerson, P. (2015). 'Human Nature'. In: Brockman, J. *This Idea Must Die: Scientific Theories That Are Blocking Progress*. New York: Harper Perennial.

Schaik, C. van & Michel, K. (2016). *The Good Book of Human Nature: An Evolutionary Reading of the Bible*. New York: Basic Books.

Schumacher, E.F. (1973). *Small Is Beautiful*. London: Blond & Briggs.

Smith, E.E. (2017). *The Power of Meaning*. New York: Random House.

Stikker, A. (2012). *En de mens speelt met de tijd, drie vensters op de eeuwigheid*. Amsterdam: Wereldbibliotheek. (Dutch)

Taleb, N. N. (2012). *Antifragile: Things That Gain from Disorder*. New York: Random House.

Toynbee, A. J. (1963-1961) *A Study of History*. Oxford: Oxford University Press.

Visser, F. (2003). *Ken Wilber: Thought as Passion*. Albany (NY): State University of New York Press.

Wilber, K. (2000). *A Brief History of Everything*. New York: Random House.

Zohar, D. and Marshall, I. (2004). *Spiritual Capital: Wealth We Can Live By*. San Francisco: Berrett-Koehler Publishers.

COLOFON

© 2018 Cees Buisman

Translation *Jetske van der Harst*
Layout *Julien Hoekstra-Kermans*

First published as *De mens is geen plaag* (2018)
Second edition 2019

Bornmeer is an imprint of
20 Leafdesdichten en in liet fan wanhoop bv

www.bornmeer.nl

IWA Publishing's authorised EU representative for General Product Safety Regulations is Diane D'Arras, 15 rue Duret, 75116 Paris, France, e-mail: safety@iwap.co.uk.

Printed and bound by CPI Group (UK) Ltd, Croydon, CR0 4YY

06/07/2026

02157567-0001